# A BRIEF HISTORY OF ERIE COLORADO

## *Out of the Coal Dust*

JAMES B. STULL

THE History PRESS

Published by The History Press
Charleston, SC 29403
www.historypress.net



*Front cover, top*: The Erie Methodist Episcopal Church was built in 1888 by James L. Wilson. The Episcopal building was moved up Holbrook Street years later to serve as the social hall and, eventually, Erie's community center. *Erie Historical Society*. *Back cover, middle*: With its Western Victorian charm, the Rose Café at 554 Briggs Street is still a popular gathering place today. *Author's collection.*

First published 2015

Manufactured in the United States

ISBN 978.1.46711.811.8

Library of Congress Control Number: 2015937992

*To Dr. Sarah Allene Wise*

# CONTENTS

# ACKNOWLEDGEMENTS

Thanks to the following people for their invaluable assistance with this project: Margaret Trostell Adams, Marianna Amicarella, Cammie Arneson, Patty Bray, Pat Bundy, Elle Cabbage, Reverend Jim Calhoun, Liz Campbell, Jess Castro, Tom Chrisman, Harold Conroe, Artie Crisp, Colleen Waneka Dame, Jim Dieke, Mary Dieke, Dennis Drumm, David Ferguson, Dean Ferguson, Patty Frobisher, John Garcia, Gayle Hepp, J.P. Holloway, Deb Jenkins, Karen Jones, Debbie Langerak, Sue Larson, Josh Lopez, Carolyn Allan McCarthy, Charlie McCarthy, Pam Mellskog, Steph Meyers, Josie Mosnik, Ruth Mosnik, Nancy Parker, Bill Pedrick, Jan Rice, Tommy Rizzi, Hallie Sawyer, James Sites, Katie Stitely, Penny Stull, David S. "Sully" Sullivan, Carol Taylor, Crystal Thornton, Chuck Waneka, Lois Regnier Waneka, Marlene Waneka, Gary Wetzbarger, Dianna Wilson, Ray Wilson, Alan Wise and Dr. Sarah Allene Wise.

You may have questions about this book's content. Many topics can be researched further by looking at the references listed at the end of this book. Erie Town Hall has records that provide minute detail. The *Images of America: Erie* book contains 213 photographs. If you believe you have more accurate information, please feel free to contact the author at eriecoaldust@yahoo.com.

I

# AN UNINHABITABLE LAND

The Town of Erie, Colorado, is located approximately twenty miles north of Denver and twelve miles east of Boulder. Before there was an Erie, the area was a barren high plain that had gone through many transformations over billions of years. Geologists have been able to identify the ages of rocks and fossils in Colorado from the Precambrian era (3.6 billion years ago) through the Cenozoic era (around 3 million years ago). They have chronicled some of the major evolutionary changes in the land—movement of continental plates, volcanic activity, the thrusting of mountains through the earth's surface, subsidence, floods, icebergs, high winds and erosion—all of which have contributed to Colorado's landscape.

More than once, oceans covered the area from the Appalachian Mountains to the Continental Divide. As recently as 100 million to 70 million years ago, Colorado was covered by six hundred feet of ocean. At other times, the land was flat and tropical, with one hundred inches of rainfall per year. The area around Denver International Airport was lush and covered with palm trees. The tilted sandstone of the Front Range was sand on an ancient beach.

About 175 million years ago, the North American tectonic plate was shifting westerly. The Pacific plate was moving northeasterly. As the plates separated, they allowed softer, more viscous matter to push up from the earth's mantle to form the Rocky Mountains. Between 85 and 70 million years ago, a subsequent shift pushed the Front Range and foothills up through the sea.

The plains are largely sedimentary layers of sandstone, shale and limestone, some resulting from alluvial deposits from rivers rushing out of the Rocky

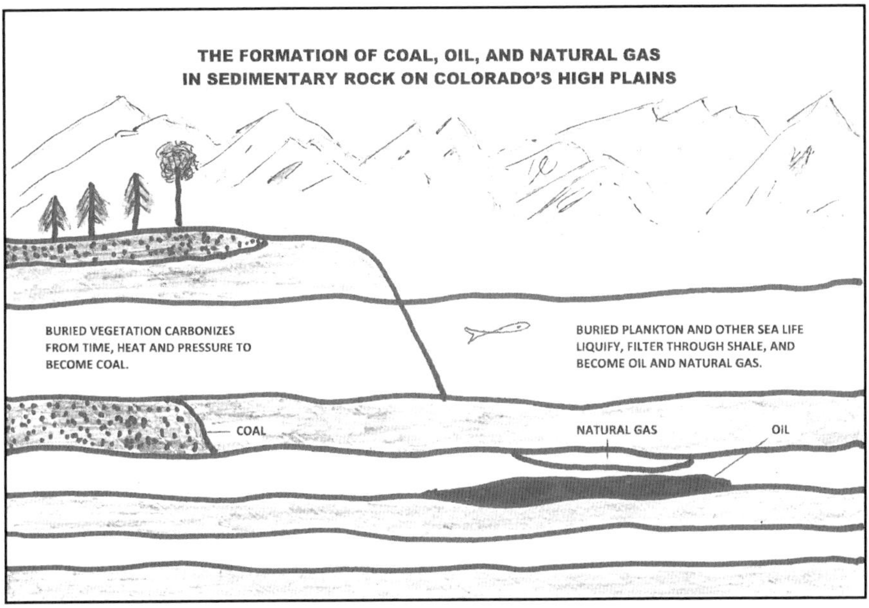


Nature provided the high plains with resources that played an important role in the establishment of the Town of Erie and in its continuing development. *Author's drawing.*

Mountains and others from the dried-up seabeds during the Cretaceous era. As these layers stacked on each other, they often trapped organisms in pockets. Trapped vegetation carbonized and became coal, while plankton and other ocean life became oil and natural gas.

The quality or hardness of coal depended on how deep and long the vegetation was buried. In the Laramie geologic formation, or Northern Colorado Coal Field, lignite and subbituminous coal was more superficial. Extracting lignite coal from the ground would provide jobs for many during the 1800s and 1900s.

Oil is made up of plankton, algae and other sea life that lived in oceans and gathered energy from sunlight. When they died, they sank to the ocean floor and were eventually covered by sediment. The former sea life liquefied and seeped through the rock, settling as reservoirs. Some of the material is extracted as oil, some as natural gas. Colorado's first oil well was drilled in the 1800s. Oil and natural gas have contributed significantly to Colorado's economy from the mid-1900s to the present.

Rivers flowed out of the Rocky Mountains, but they only reached narrow strips of land throughout the Great Plains. After the oceans disappeared, the area was dry, sandy and bare of vegetation. Because of alluvial deposits from the rivers, the ground was full of gravel, with no topsoil for cultivation. It provided neither natural shelter nor trees for eventual visitors to build houses. The temperatures ranged from over one hundred degrees in the summer to negative thirty degrees or more in the winter. Torrential erosion, dust storms, hail, severe winter blizzards, tornadoes and grasshopper and locust infestations further contributed to making the Great Plains inhospitable.

Coal Creek flows near Old Town Erie. About twenty-nine miles long, it originates in Coal Creek Canyon on the eastern flank of Thorodin Mountain near Wondervu in the Front Range. It travels through Superior, Louisville and Lafayette. It enters Weld County at the southwest tip of Erie, runs east of Kattell Street and leaves Erie at Highway 52 to merge with Boulder Creek. The water then follows the course of the St. Vrain River, South Platte River, Platte River, Missouri River, Mississippi River and into the Gulf of Mexico.

Although wildlife and nomadic Native American tribes probably drank from it, Coal Creek was never the source of Erie's drinking water. By 1858, it had become contaminated by various toxic chemicals used to expose minerals extracted by gold and silver miners in the Rocky Mountains. Still today, large volumes of cadmium, copper, manganese, lead and zinc are discharged from the mines and pollute many of Colorado's rivers and streams, making them devoid of fish and other aquatic life and unfit for human consumption. However, creek water and mine wastewater were used for washing laundry and other purposes.

The first recorded flood swept through Erie in May 1876. Railroad tracks were washed out and had to be replaced. In 1890, another flood devastated parts of Erie near the creek. In 1920, a dike was built along Kattell Street. In 1921, Erie suffered the worst flood in its history when the dike broke. Some houses floated off their foundations. Bridges and railroad tracks were washed away. Reportedly, water was as high as countertops in the stores on Briggs Street. After cleanup, a wider channel was dug. In 1930, another dike was built along Kattell Street, but it gave way during subsequent flooding. In 1969, Coal Creek flooded again. Schools closed so students could help sandbag the creek. Bridges washed out and roads were closed. In 1972, another flood damaged a bridge north of town on County Line Road and the Coal Creek Bridge east of town. A new dike was constructed. In 2013, Erie suffered another devastating flood that also impacted many nearby towns and counties.

Today, Coal Creek seems small and nonthreatening; however, Erie still faces the potential for occasional flooding. Town officials say digging and drainage projects have mitigated damages from future rainstorms. Because most of the efforts focused primarily on the Old Town section of Erie, more outlying areas are still at risk.

Where the first inhabitants of North America came from is a debated topic. Based on linguistics, genetics and blood types, a popular theory is that the first people on the continent crossed the Bering Strait from Asia some twelve thousand to forty thousand years ago, following food sources and seeking acceptable places to settle. This theory suggests these visitors were the ancestors of Native Americans in North, Central and South America.

Archeologists have uncovered evidence of travelers from the Bering Strait crossing along the Front Range from Wyoming to New Mexico. After a flood of the Big Thompson and Little Thompson Rivers in 1932, a Union Pacific Railroad employee discovered large mammoth bones and projectile points dating back some eleven thousand years. The findings are believed to be from the Clovis culture. While many cultures drifted through the area we call Erie, there is no evidence any of them ever settled there.

Several tribes of Native Americans visited Colorado: Sioux (Dakota, Lakota, Nakota), Pawnee, Ute, Pueblo, Kiowa, Apache, Comanche, Crow, Shoshoni and Blackfeet. Many have been identified as having visited Weld County.

Historians estimate that around 1820, the Arapaho and Cheyenne tribes left Minnesota and North Dakota, respectively, and traveled together toward the west. They felt pressure from the influx of white settlers and the flexing of muscles by the Sioux. The Arapaho and Cheyenne came from the Algonquin language group. Their individual languages differed, but they were able to communicate and got along well while they searched for a new place to live. While they maintained their individual tribal customs, they did intermarry. Around the mid-1850s, the approximate time white visitors came into the Erie area, so did the Arapaho and Cheyenne. Historians are uncertain about who arrived first. The tribes were nomadic, moving their camps based on weather and the availability of food. The area provided deer, elk, antelope and buffalo and was fairly pleasant during the winter. Tribe members lived in tipis and mold huts; some may have lived in caves and dugouts along Coal Creek, where large numbers of arrowheads have been found. Library archives show pictures of Arapaho and Cheyenne bathing in the creeks near Erie.

Explorers were commissioned to find trails through the Rocky Mountains and Sierra Madres to the Pacific Ocean, among them Zebulon Pike, Stephen H. Long, John C. Fremont and Stephen W. Kearney. Pioneers crossed the plains in search of new homelands in the West. Those seeking their fortunes during the California and Pike's Peak gold rushes in 1849 and 1858 hurried

Chief Hosa (circa 1810–1869), also known as Little Raven and Young Crow, was a major chief of the Southern Arapaho. *National Archives.*

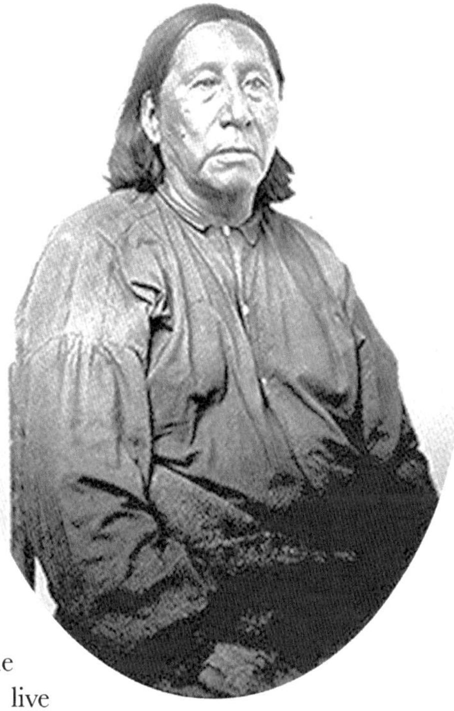

through the area. There is no evidence that any of these groups settled near Erie. People questioned whether the plains had the potential to be fertile agricultural land. About the only thing that grew in abundance in this arid region was a resilient buffalo grass. They deemed the area uninhabitable.

In 1857, the United States went through a recession that brought on hard times for many of the nation's workers. People began to look for new places to live that were less crowded, not industrialized and had better air and water quality. Others came west to escape the impending Civil War.

The earliest known white visitor to the Erie area was Jim Baker. Jim was born in Bellville, Illinois, in 1818 into a family of Scotch Irish descent. Around 1838, he walked to St. Louis to sign on with the American Fur Company, a group that sent trappers west to bring back valuable beaver pelts. Jim Bridger, a well-known trapper, showed Baker how to survive in the trade. Baker was somewhat of a latecomer, but he quickly mastered the skills. Eventually, he went off on his own, realizing he could make more money if he didn't have to share the profits from the sale of his pelts. Among the many people Jim befriended during these years was Kit Carson.

From 1838 to 1839, Baker hunted and trapped in the Wind River Mountains in Wyoming. In 1840, he returned to Illinois, but he quickly realized he wanted to continue trapping. He went back to Wyoming. In August 1841, he and about two dozen other trappers engaged in a fierce battle with members of the Sioux, Cheyenne and Arapaho tribes at the junction of Bitter Creek and the Little Snake River. The battle site was near an extinct volcano formerly called Bastion Mountain. The area is known today as Battle Mountain.

In 1845, Baker rode with General John Fremont on an expedition to the West. They found a route to Los Angeles. On another visit, Jim and others returned to California to steal thousands of horses for General Fremont.

In 1846, Baker was welcomed into a Shoshoni camp, where he met and married Marina, the daughter of Chief Washakie. Jim became known as the redheaded Shoshoni. Sources vary, but Jim is reported to have married at least six Native American women, mostly Shoshoni, one Cherokee and one Sioux. He fathered eleven children. Because one of his wives was a princess, Jim was regarded as a chief.

In 1847, Baker settled in Salt Lake City. In 1857, he was assigned to fight against the Mormons when the United States government objected to Utah wanting to secede from the United States under Brigham Young's leadership.

A rite of passage for mountain men was to fight a grizzly bear. Allegedly, Jim and a partner encountered two bears one day, and Jim killed both bears after his partner was injured. Jim lost part of his face in the fight. At another point in his life, he lost more of his face when his rifle misfired. The good looks he may have enjoyed as a young man changed. One child told his father that Jim Baker was the ugliest man he had ever seen.

Jim came to Erie around 1858, camping on the west bank of Coal Creek. He is credited with having owned the first coal mine in Colorado. He called this slope mine Baker's Bank. It, too, was located along Coal Creek. He occasionally drove a wagonload of coal to Denver to be sold to business operators; however, selling it in Denver at one dollar per bushel (eighty pounds) or twenty-five dollars per ton proved to be unprofitable so he left. Jim lived in Denver for a time near what is now Seventeenth and Tennyson Streets, near Sloan Lake Park. He owned a bridge over Clear Creek, known as Baker's Crossing. One of his wives collected tolls from those wanting to cross the creek.

In 1859, Jim was commissioned as a captain in the United States Army. It is believed that John Chivington was promoted in the same ceremony. Chivington led the charge at the Sand Creek Massacre in 1864. Because he spoke fluent Shoshoni and knew Arapaho sign language, Baker served as an Indian scout for the Colorado state militia and federal troops. He rode with General George Crook at the Battle of the Rosebud in 1876, gathering intelligence on the Lakota for General George Custer. The Rosebud battle showed that the Lakota were both excellent horsemen and good fighters.

Allegedly, Jim's scouting report never reached General Custer, and the lack of his findings is considered to be a major factor in Custer's defeat at Little Big Horn. Baker rode with General Thornburgh at the Meeker Massacre in 1879, where the United States Army battled with the Ute tribe over land rights. The result of this skirmish for Baker was a long-lasting bounty on his life by the Ute.

Jim died on May 15, 1898, in the two-story cabin he built near Savery, Wyoming. It had been a three-story cabin during times when the various Native American tribes were unfriendly, with the upper story being used as a watchtower. The tower was destroyed when the cabin was moved to Cheyenne to be displayed in a museum. Today, his cabin may be visited at the Little Snake River Museum in Savery.

A few families farmed land near the creeks, and merchants provided goods for those farmers, but the land was still virtually uninhabited. In 1862, Congress passed the Homestead Act, authorizing the United States government to give away 160-acre parcels of land to anyone seriously intending to live on the land and cultivate it. All one had to do to obtain title to the land in perpetuity was to build a twelve- by fourteen-foot habitat on it and farm forty acres for at least five years. This act officially went into effect in 1863 and contributed to the early growth and greatness of Colorado.

As white settlers later homesteaded the West, much of the wild game left the area. The Arapaho and Cheyenne were not welcome and were driven away from the homesteaders' properties. Eventually, the tribes were assigned to a reservation in Kiowa County called Sand Creek. There was a movement to rid Colorado's plains of all Native Americans. At sunrise on November 29, 1864, several hundred volunteers and soldiers attacked the Sand Creek settlement, slaughtering most of the tribe members and ordering the rest to move on to other reservations. Today, most Native Americans in Colorado are part of mainstream society. Those Arapaho and Cheyenne who remained on reservations live in Wyoming and Oklahoma, some being forced to live with tribes that were previously their enemies.

In the late 1860s, miners from the eastern United States, Great Britain and Germany learned of the abundance of coal in Colorado. This led to more people coming to the Erie area to find a better lifestyle.

2

# EARLY SETTLERS

The earliest settlers were farmers who were in the area before the Homestead Act of 1862; they grew wheat, corn, alfalfa, barley, oats, sugar beets, cabbages, turnips, carrots, cucumbers, lettuce and a variety of fruits. Some raised horses, cattle, hogs, sheep, goats, chicken, turkeys and ducks. Once the opportunities associated with the Homestead Act were publicized, people flocked to Colorado. Some of the early homesteaders in the Erie area included the Wise, Plumb, Daily, Smith, Carr, Liggett, Marfell, Pease, Kempton, Leyner, Hauck, Prince, Tyler, Delechant, Beasley and Miller families.

When coal was discovered near Erie in 1866, miners arrived from Europe, Mexico, Japan and the eastern United States. This raised the need for boardinghouses, saloons, supplies stores, livery stables, blacksmiths and other businesses necessary for a coal mining town. Miners had to provide their own shovels, picks, augers, breastplates and other equipment. As the miners' families eventually followed them to Erie, a wider variety of stores and services became necessary.

When people arrived, they tended to seek out other residents with similar cultural backgrounds. They often lived near each other in segregated sections of town to escape prejudice. A two-block area at the north end of Briggs Street became known as French Town; however, because the French were continually isolated from the rest of the community, they eventually left to live in other towns where they felt more comfortable and appreciated. Greeks lived in an area toward the south end of Erie known as Goat Hill.

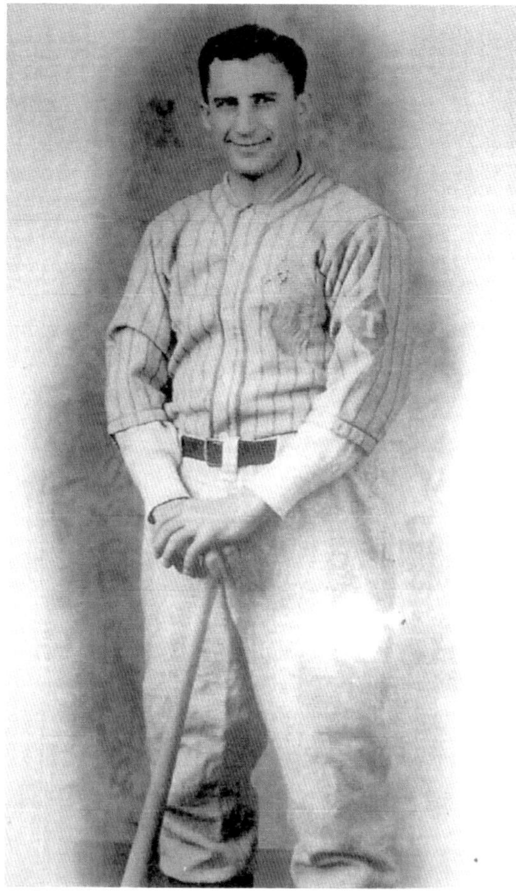

Henry Amicarella (1915–2011) was inducted into the University of Northern Colorado Hall of Fame for his football achievements and was an All Rocky Mountain Conference baseball shortstop. *Marianna Amicarella collection.*

It is impossible to list all of Erie's earliest settlers; however, available information on a few is worth sharing. Early residents are also listed in other chapters, covering Erie's mayors, the Coal Miners' Memorial and historic street names. Erie's Town Hall records office has documents with many more names.

**Henry Amicarella** was born in 1915 and raised in a coal mining camp near Glenwood Springs. In 1922, his family moved to Serene near Erie. In 1927, he watched bullets penetrate the walls of the family cabin as the militia fired weapons at striking miners, resulting in the Columbine Mine massacre. Just twelve years old, he knew he didn't want to be a miner. After graduating from Erie High School, he attended college in Greeley. He attended graduate school, earning two master's degrees, one from Northwestern University. He became a hospital administrator in Longmont. After retirement, he volunteered to take health care assistance to numerous foreign countries and organized the Tiny Tim Center and Meals on Wheels. He was recognized as an honored alumnus by Erie High School in 2003 and was named citizen of the week. He died in Longmont in 2011.

**Ira Austin** was born in New York in 1801. After he moved to Colorado, he traded his flock of sheep for a parcel of land owned by Henry Gilson

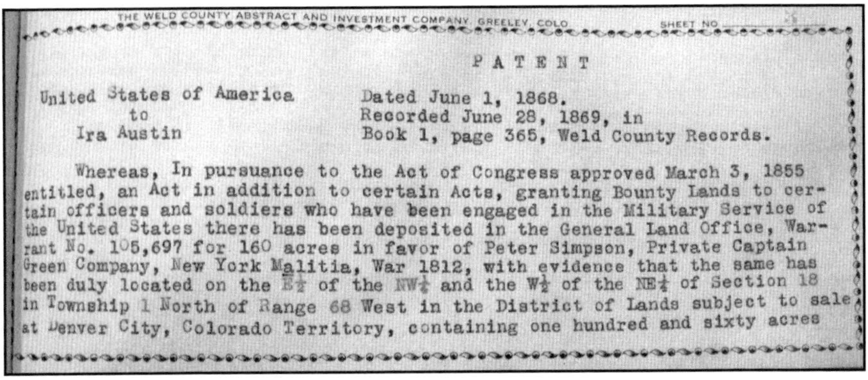

THE WELD COUNTY ABSTRACT AND INVESTMENT COMPANY, GREELEY, COLO. SHEET NO.

P A T E N T

United States of America
to
Ira Austin

Dated June 1, 1868.
Recorded June 28, 1869, in
Book 1, page 365, Weld County Records.

Whereas, In pursuance to the Act of Congress approved March 3, 1855 entitled, an Act in addition to certain Acts, granting Bounty Lands to certain officers and soldiers who have been engaged in the Military Service of the United States there has been deposited in the General Land Office, Warrant No. 105,697 for 160 acres in favor of Peter Simpson, Private Captain Green Company, New York Malitia, War 1812, with evidence that the same has been duly located on the E½ of the NW¼ and the W½ of the NE¼ of Section 18 in Township 1 North of Range 68 West in the District of Lands subject to sale at Denver City, Colorado Territory, containing one hundred and sixty acres

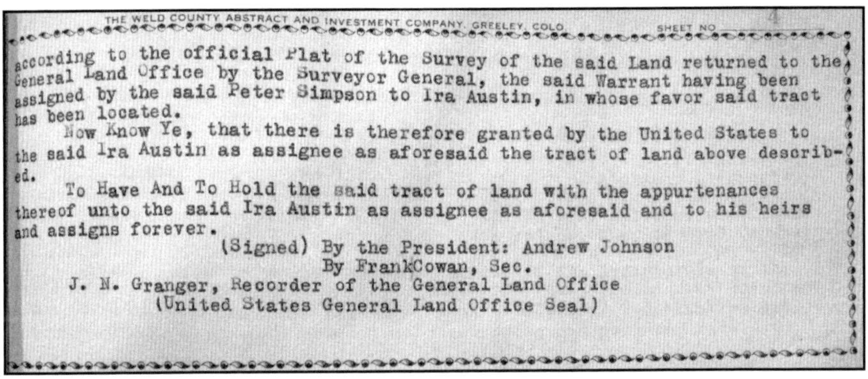

THE WELD COUNTY ABSTRACT AND INVESTMENT COMPANY, GREELEY, COLO. SHEET NO. 4

according to the official Plat of the Survey of the said Land returned to the General Land Office by the Surveyor General, the said Warrant having been assigned by the said Peter Simpson to Ira Austin, in whose favor said tract has been located.

Now Know Ye, that there is therefore granted by the United States to the said Ira Austin as assignee as aforesaid the tract of land above described.

To Have And To Hold the said tract of land with the appurtenances thereof unto the said Ira Austin as assignee as aforesaid and to his heirs and assigns forever.

(Signed) By the President: Andrew Johnson
By FrankCowan, Sec.
J. N. Granger, Recorder of the General Land Office
(United States General Land Office Seal)

President Andrew Johnson signed Ira Austin's patent for land in Erie in 1868. Weld County recorded an abstract of the document. *Karen Jones collection.*

just east of Erie; he discovered a large coal vein on that property in 1866. In 1868, Ira purchased a 160-acre parcel of land where the original town of Erie would exist; he purchased it through a "scrip" patent, available under the Congressional Acts of 1842 and 1855 to those who performed military service. He and his son, Schuyler, opened the first official coal mine there in 1876. Ira was also a miner and a farmer and lived in Boulder in the 1880s.

**George Bain** and his wife came to the United States from Scotland in 1849. They eventually settled in Erie. Photographs of a buckboard wagon with Bain's name on it suggest that he was a merchant.

**Ransom Balcolm** was a New York Supreme Court justice who traveled to Colorado and helped organize the Boulder Valley Mining Company. He was also instrumental in platting the Town of Erie. There is no evidence that he ever lived in Erie, but Balcolm Street is named after him.

**Dr. Clarence W. Bixler** was born in Ohio in 1877, where he attended Ohio State University. He relocated to Boulder to attend the University of Colorado medical school. He graduated in 1907 and moved to Erie with his wife and two daughters to set up his medical practice. He had limited funds, so he leased a horse and cart from a livery stable so he could make his rounds, attending mostly to families of farmers and miners. He served the tri-town area of Dacono, Frederick and Firestone, at a rate of fifty cents per patient. He was offered salary from Erie and Lafayette mining companies and was able to maintain a better living standard. In 1909, he bought a house at 485 Holbrook Street. Joe McWilliams's wife, Anna Zimmerman, served as his nurse in 1927, when Dr. Bixler treated and saved the lives of several miners who were wounded by the militia during the Columbine Mine massacre. He is said to be largely responsible for the introduction of penicillin and insulin to the Boulder County area. He also purchased the first X-ray machine in Colorado. This technology drew miners and others to his practice. He was a civil servant, president of Erie Bank and a champion for changing the boundaries of Erie and moving the entire town into Boulder County. In 1936, he moved to Longmont to continue practicing medicine there. He loved agriculture, and he purchased thousands of acres of prime farmland in Weld and Boulder Counties. He died at home in Longmont in 1949 at the age of seventy-two. In 1969, his daughter, Ruth, sold some of her father's farmland and started the C.W. Bixler Family Foundation. In 2006, she donated $9 million to the University of Colorado, part going to the Boulder campus to be used for the art, art history and music programs and the rest going to the Denver campus for its agriculture program. Clarence Bixler was one of the most beloved and well-known doctors in the state of Colorado.

**Daisy Swallow Bracegirdle** was one of the early owners of a telephone in Erie and served as a telephone operator. She often was the first to learn of trouble at the mines in Louisville, Lafayette and Marshall. She was sympathetic to the causes of the miners. Clarence Darrow slipped into town and gave Daisy coaching on how to testify in court when she was required to appear regarding her knowledge of anything related to the Hecla Mine shootings. When asked if she knew if William Knowles had talked to John Lawson or Edward Doyle the night of the incident, she said it wasn't her place to listen in on private conversations. Co-defendants in the case were her cousin Arthur Spendlow and her brother-in-law Gus Brack.

**Charles Brannan** was the secretary of agriculture under Presidents Harry S. Truman and Dwight David Eisenhower. While he was born and died in Denver, one source listed him as affiliated with Erie through his legal expertise in mining and irrigation.

**DeWitt Brennan** was the great-great-grandson of John D. Williams and Mary Phillips Evans. As a youth (circa 1920), he sold popcorn at the Erie Opera House.

**James Brennan** was president of Erie Bank and a member of the Anti-Saloon League in 1907.

**Mike Brennan** was part owner of the Star Mine along with William O. Wise.

**Henry Briggs** was born around 1828 in New York. He owned and operated Erie's first commercial coal mine, the Briggs Mine. Eventually, he was an official of the Boulder Valley Mining Company. The Briggs Mine was later known as the Boulder Valley Mine. Records suggest that in 1882 he was involved in legal proceedings with the Denver Western and Pacific Railway. In 1868, Henry purchased the eastern quarter (160 acres) of the square-mile section of land that made up most of Erie's original boundaries. His parcel was located primarily east of Coal Creek. He purchased it through a "cash-entry" patent.

**Isaac "Ike" Canfield** and his brothers struck a coal seam, which they called Rob Roy. It later became known as the Jackson Mine. Ike was also one of the eleven people named on the November 11, 1878 incorporation papers for the "Baby Railroad," the narrow gauge train that ran from Canfield to Longmont. He went into the oil speculation business in the Boulder area but found the cost to capture the oil prohibitive. After gaining more experience, he did locate some oil near Niwot. Canfield was named after him.

**Joseph Cook** was born in England in 1817. His father was Sampson Cook. When Joseph was twelve, he accompanied his family to the United States. He married Elizabeth Davey in 1877 and moved to Erie in 1878. He worked in the Clayton and Garfield mines. His brother, George, was injured in a Lehigh Mine accident.

**Joseph Cundall** was one of Canfield's early pioneers. He was born in England in 1874. He married Lucy Tonse; they were the parents of eight children.

**William A. Davidson** was educated at the Phillips-Andover Preparatory School. He founded the Davidson Coal and Iron Company. He later joined forces with officers from the Colorado Central Railroad. Davidson planted many acres of sugar beets where miners often worked during the summers. He also built a $13,000 water canal from South Boulder Creek to the coal fields. He was involved with various other ventures.

**William Henry Drinkwater** was born in 1858 and became co-owner of the Garfield Mine with William Whiles. Harriett Drinkwater is listed in the Erie cemetery as born in 1855 and assumed to be William's wife. Harriet died in 1929; William died in 1933.

**Charles Elzi** was born in Boulder in 1890. He came to Erie in 1912, directly from Denver, to open a drugstore; he had earned his pharmacy degree from Capitol College. His drugstore at 502 Briggs Street had a soda fountain with a mirrored back bar. It was one of the earliest licensed drugstores in Colorado and one of the first to dispense insulin. Elzi served as mayor from 1940 to 1942 and 1944 to 1946. He was an active charter member of the Lions Club; people say he never missed a meeting. He was also president of the Erie Bank. He married Doris Wilby in 1917, and they had four children—William, Charles, Betty and George. Doris was a teacher in Erie. Charles died in 1972 and is remembered as being helpful and friendly to the community.

**Cornelius Enders** was a German-born Catholic priest who started St. Scholastica Church. He and his parishioners—primarily Slav, French, Lithuanian and Italian miners—hauled lumber from the sawmills in the mountains and worked together to build the church.

**William "Bill" Ferguson** was the son of Scottish immigrants who first moved to Illinois to work in the coal mines. They moved to Colorado, where Bill continued his lengthy career. At the age of seventeen, he began working as a mule driver. He worked at many mines, leaving one to take a better job at another. The mines he worked at include the Monarch, Industrial, Columbine, Lincoln, Washington, Morrison (new and old), Acme, Grant and Liley. Bill and his family lived in some of the mining camps near Erie. He worked for the Rocky Mountain Fuel Company as well. He ended his career in 1972 at age seventy-two as president of the Clayton Coal Company. He died at the age of ninety-two.

**Imo Miller Fletcher** moved to Erie in 1921, with her parents and siblings, so they could be closer to Dr. Clarence Bixler, who had been caring for Imo's sister. Imo married Joseph "Cotton" Fletcher, a miner.

**Joseph "Cotton" Fletcher** worked at the Washington Mine from the time it opened until it closed in 1967.

**C.P. French** was Erie's first physician, in 1875. By 1885, Erie had four doctors. By 1890, only three doctors remained in Erie.

**George Gilson** discovered coal on his parcel of land around 1859. This was the first official discovery of coal in Erie. In the mid-1860s, he traded the land for a flock of sheep owned by Captain Ira Austin.

**Paul Gorce** was born in 1878 in France and killed on September 17, 1915, in Benjamin's Saloon in Erie. He had broken up an argument at the bar in the afternoon, and when he returned in the evening, one of the men he confronted earlier drew his gun and killed Gorce. Paul's brother, Martin, was also shot in the arm, but not before he knocked out the shooter with a baseball bat. Martin was the brother-in-law of Sheriff Sanford Buster.

**Steve Gorce** immigrated to the area from France in 1879. He worked at the Puritan Mine. He was the father of Paul and Martin Gorce. Steve was regarded as a leader who had gained the trust of strikebreaking miners. He provided them with a good orientation about how to proceed during the strikes. He was deputized by Sheriff Sanford Buster because of his rapport with the miners.

**Julia Gordon** was a midwife and a practical nurse. Midwives would often stay at the homes of pregnant women for several days before and after the actual birth. Midwives frequently worked for free.

**John Grimson** was born in Scotland in 1866 and came to Erie with his mother and stepfather around 1880. John and his wife, Mary Webber, had eight children. John was rescued from a cave-in at the Lehigh Mine but died in 1907 or 1908 from injuries suffered in a second accident.

**Mary Webber Grimson** came to Erie from Wales in 1879. She was the daughter of James and Gwendolyn Webber; a conflicting report lists her parents as Matthew and Margaret (Reynolds) Webber. Mary's three brothers—John, Thomas and William—were killed in an explosion at the Schofield Mine in Utah in 1900. Following her husband's death, Mary wed Harry Wright Gregg. The Gregg family sold their land to the school district in 1926, and it became the home of the Erie Junior/Senior High School on Main Street.

**Edward C. Kattell** was a Tioga County judge from New York. He was also president of the Binghamton (New York) Oil Refinery. With Ransom Balcolm, he organized the Boulder Valley Coal Company and helped plat the town of Erie. Kattell Street is named after him. He is not believed to have resided in Erie.

**William "Bill" Knowles** immigrated to the United States from England with his wife and three children in 1861. He was a farmer and a miner, and eventually, he was president of Erie local of the United Mine Workers Association (UMWA). He also served as mine superintendent, justice of the peace and a member of Erie's town board. His father was killed in a mine in the Northern Coal Field. At fifty-four, William was tried for attempted murder following the Hecla Mine incident. He was convicted of the charges. He died in 1926 and is buried in Erie's Mount Pleasant Cemetery.

**Harold "Harry" Larson** was named Harald Gottfred Larsson at birth on November 19, 1895, in Hanna, Wyoming. He met Fred Wilson in Hanna, when Fred went there from Erie to learn how to run a store for the Union Pacific Railroad. He also met and married Jean Wilson, who had gone to Hanna from Erie to teach school. Harry and Jean were married in 1921. They moved to Erie shortly after that. Harry and Fred co-owned the Wilson-Larson General Store, which was located in the two-story building at 500 Briggs Street.

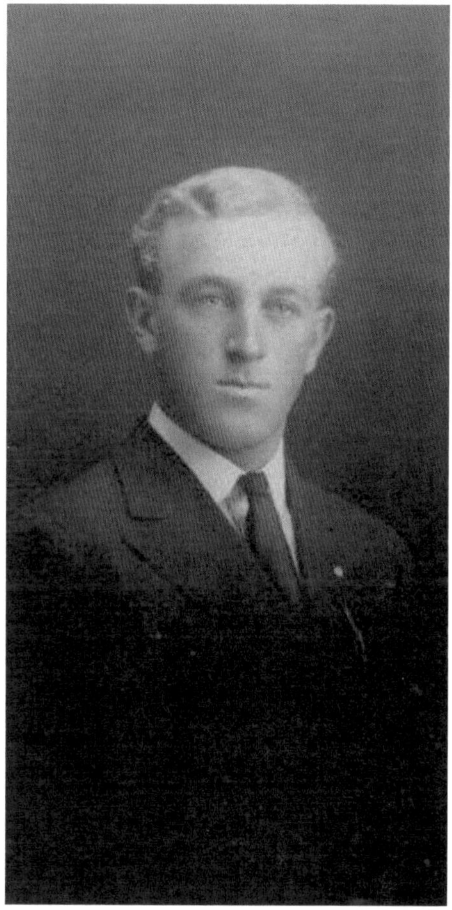
Harry Larson co-owned and operated the Wilson-Larson General Store from 1921 until the 1950s. *Sue Larson collection.*

**George Laughlin** was a miner in Erie until 1905, when he moved to Frederick and started a livery stable. His daughter was Eva Fernandez.

**Stella Lee** was the daughter of James L. and Charlotte (Richards) Wilson. She and her siblings—Fred, Jean and Francis—were born in Erie. Stella was a schoolteacher in Erie and at the Columbine Mine. She married Jay Lee, and they had one son, Robert W. Lee.

**Ella Leyner** had great skills at designing, making and trimming ladies' hats. She made a living as a milliner.

**Sabino "Sam" Lontine** (also spelled Lontin and Lonteen) was born on October 30, 1896, in Hastings, Colorado. He worked as a miner at the Puritan Mine. He also served in the U.S. Army in World War I. He and his wife, Dora Muniz, raised ten children: David Roger, Betty, Charles Julian, Sabino "Sonny," Leroy, Franklin, Mabel, John Lee, Everett and Ken. Erie High School principal Harold Conroe reports that the Lontine children were normal and loved to have fun, but they were always well behaved in school because they knew the consequences if they were not. Sam died on September 29, 1989, in Longmont.

**Hiram Marfell** was born in England in 1840. He immigrated to the United States to work in the coal mines; he worked in the Mitchell Mine near Erie. He was also an original homesteader. He married Eliza Flanagan.

**James Marshall**, John's son, opened a drilling business around 1916. His company drilled test holes to determine the potential value of various mines in the area. The company also drilled many water wells. He and his wife had three children—Inez, Josephine and James. Josephine died at seventeen during an influenza epidemic.

**John Marshall** was born in Scotland and moved to Erie with his family in 1883. He lived near the Boulder-Weld County Line Road with his wife, two sons and six daughters. He negotiated with the Burlington Railroad because they wanted an easement through his property. His land also contained the Boulder irrigation ditch, which became part of the Big Thompson distribution system. John set up a livery stable near Briggs and Moffat Streets. He also owned a drugstore building and a house on Briggs Street.

**Ida McAfee** was born in Erie and married William Waneka in 1897 or 1900 (accounts differ).

**Virgie McAnally** left Erie around 1916 to become postmaster in Frederick.

**Walter McAnally** owned the first print shop in Erie and was the first publisher of the *Erie Herald*. He had a daughter, Pauline.

**John McCool** was granted the first patent for coal mining in Erie, signed by President Ulysses S. Grant on April 5, 1871. His 160-acre parcel was the most western of the square-mile section he shared with Henry Briggs, Ira Austin and Schuyler Austin. John was a boot and shoe merchant from Denver.

**W.E. McCoy** co-organized sixty miners and formed Local 711 of the Knights of Labor, the first national union in Colorado.

**Gertrude Morgan McDonald** moved to Erie with her family in 1904. She was married to John McDonald. She was a store clerk and a postal clerk. Her father owned the Shamrock, Imperial and Eagle Mines. She had one son, Dr. John "Jack" McDonald.

**John McDonald** was a barber and justice of the peace in Erie for twenty-five years. He also served as postmaster and an insurance agent. He married Gertrude Morgan.

**Joe McWilliams** managed the commissary at the Puritan Mine camp. He and his wife, Anna Zimmerman, moved to Erie to open a meat market on Briggs Street. He eventually lost the store during the Great Depression.

**George Meller** was present at Erie's incorporation in Greeley in 1874 and served as president of Erie's first board of trustees.

**Emmett Metheney** arrived in Erie in 1921 from Ohio to work in the mines. He and his wife, Belle, had four children—Mary, Ruby, Edna and Woodrow. The girls attended Erie High School. Belle, Mary and Edna became charter members of Pythian Sisters. Family members are buried in the Metheney family plot in Erie Cemetery.

**Chris Miller** was a baker who supplied biscuits for Erie Biscuit Day during the late 1800s.

**C.C. Montgomery** was a farmer. He raised four children in Erie. He remembers a flood during the 1890s and that he and his mother were trapped and couldn't return to Erie from Boulder.

**John S. Morgan** was a civic leader and merchant, born in 1853 in Pennsylvania; his parents had emigrated from Wales. He erected a large Queen Anne–style residence with brick coverings at 704 Main Street. He served as postmaster and as a trustee in 1879. He operated the Morgan Brothers general merchandise store with his brother, J.J. Morgan. He bought out his brother's interest in the store in 1892. He moved his store into the Independent Order of Odd Fellows (IOOF) building at 500 Briggs Street in 1897. He is listed in the state business directory as a merchant until at least 1903. In 1909, he sold his house to Henry Carter and moved to Denver.

**George Morrison** was a Scottish immigrant who started as a blue-collar worker. In 1900, George became the superintendent of a mine in Superior. In the 1910s and '20s, he supervised mines in Routt County. In 1930, he became president of Clayton Coal Company. He also owned and operated the Morrison Mine, which employed between two hundred and five hundred men. Northwest of Erie, it was one of the seven largest mines in the Northern Colorado Coal Field. He erected a house in 1890 at 675 Holbrook Street. His parents were Alexander and Mary R. Morrison.

**Alvin Mosnik** was born in England. Alvin served in the United States Army during World War II. His wife, Ruth, married and joined him in Erie in 1946.

**William Nicholson** was born in 1850 in England. He began working in the mines in England at age eight. He moved to America in 1881. In 1882, he married Philadelphian Alice Pollatt. He worked in the mines in Ohio until 1883, when he moved to Erie and became a superintendent of coal mines. He invested in coal properties. In 1892, he founded the Long's Peak Coal Company and developed the Longs' Peak Coal Mine. He sold that enterprise to the Rocky Mountain Coal Company. In 1903, he founded the Erie Bank, and after retiring, he served on its board of directors. He resided at 604 High Street, which, at that time, was considered the finest residence in Erie. He served five separate terms as mayor: 1888–90, 1907–10, 1910–14, 1915–21 and 1929–30.

**Job O'Taylor** came to the United States from Great Britain to be a miner at the Parkdale Mine. He was killed in 1907 by a runaway coal car. Taylor Lane in Vista Point is named for him.

**Mary Payne** was born in Erie in 1890. Her father was D.F. Gordon, superintendent of the Long's Peak Mine and engineer of the power company that produced electricity for Erie in the early 1900s.

**Martha "Mattie" Phennah** (maiden name unknown) was born in 1860 in Wisconsin; her parents had emigrated from Wales. She was one of Erie's businesswomen. In 1884, she married Peter Phennah. He was an engineer who was born in 1855 in Wales; in different business directories, he was listed as a state engineer and a stationary engineer. Mattie opened a dry goods and millinery store. The 1900 United States census lists them as living at 684 Holbrook Street and having four children. In 1917, Michael James Brennan purchased their house. The United States census for 1920 lists them as living in Seattle, Washington.

**Pauline Phillips** played the piano at the Erie Opera House when it showed silent movies. She and her husband, Lewis (Lou), had two daughters, Lorraine and Dorothy. Lou was a coal miner. Dorothy was on the Erie High School girls' basketball team that won the Colorado state championship in 1929. Pauline was an active member of the Sharon Rebecca Lodge and the Pythians. She was the daughter of Walter McAnally. She died in 1983 and is buried in Erie's Mount Pleasant Cemetery.

**Charlotte "Lotti" Pickett** was born in 1891 at the Picton Coal Camp near Walsenburg, Colorado. Her husband, Lawrence, worked for thirty cents per hour, ten hours a day in the mine. They moved to Erie in 1921 so Lawrence could continue working as a coal miner.

**B.C. "Bud" Pitchford** served as Erie's constable/marshal and its only maintenance man; he lit the streetlights each evening, graded the streets during the summer and removed ice and snow from them in the winter. When he thought Louisville was under attack by striking miners and that Erie would be next, he is suspected of ordering the lights to be turned off in town.

**Charles Preuss**, from Germany, was Captain John C. Fremont's chief cartographer. Preuss traveled with Fremont and Lieutenant Zebulon Pike. Preuss eventually settled in Erie.

**Hiram Prince** was an original homesteader in Erie. Realizing the need for irrigation for Erie's farms, he helped divert water from Boulder Creek to

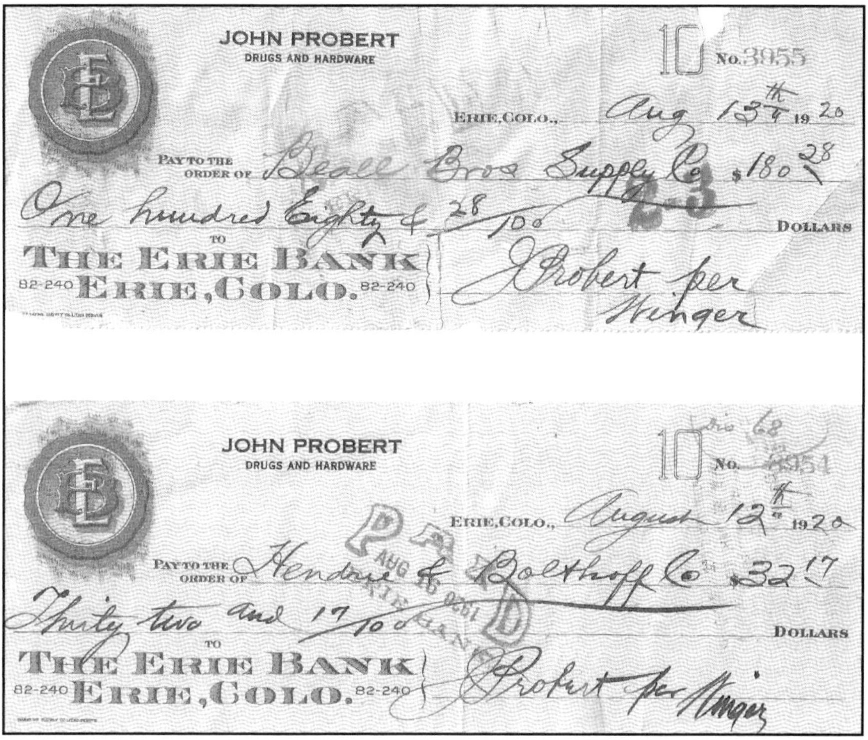
JOHN PROBERT
DRUGS AND HARDWARE
No. 3955
ERIE, COLO., Aug 13th 1920
PAY TO THE ORDER OF Beall Bros Supply Co $180 28/100
One hundred Eighty & 28/100 DOLLARS
TO THE ERIE BANK
82-240 ERIE, COLO. 82-240
J Probert per Winger

JOHN PROBERT
DRUGS AND HARDWARE
No. 3954
ERIE, COLO., August 12th 1920
PAY TO THE ORDER OF Hendrie & Bolthoff Co $32 17/100
Thirty two and 17/100 DOLLARS
TO THE ERIE BANK
82-240 ERIE, COLO. 82-240
J Probert per Winger

These checks from John Probert's store are from the 1920s. The signature lists "per Winger," who eventually operated a hardware store at Briggs and Wells Streets. *Debbie Langerak collection.*

form Erie's three irrigation reservoirs on the southwest corner of the town near today's U.S. 287 and Arapaho Road.

**John Probert** left Wales at the age of seventeen to seek out a life in America working in coal mines. After working in Ohio and Illinois, he left for Colorado to work in the newly discovered coal fields. He soon learned that he wanted to work in a field other than coal mining, and he took a job as a salesman in John T. Williams's general store. Eventually, he earned his pharmacy license, only the thirty-eighth issued by the State of Colorado. He opened Erie's first drugstore.

**Sarah Probert** served Erie women as a midwife. Midwives were necessary because doctors had to attend to so many other patients.

**Ira Reddington** moved to Erie from Lafayette in 1941. He and his wife, Dottie, owned and operated the Northport Store northeast of Erie from 1942 until 1972.

**Julius Regnier** was Mable Regnier's father-in-law. At the age of nineteen, he came to the United States from Belgium to settle in Trinidad, Colorado. He moved north to farm but quickly moved to Erie to work in the coal industry. He had worked in coal mines since he was ten years old. He retired from the Boulder Valley Coal Company in 1962.

**Mabel Regnier** was the daughter of Harriet (Pitchford) and William H. Drinkwater. She became one of Erie's telephone operators. She married Albert Regnier in 1919. They had two children, Lois Jane (Waneka) and DeVerne.

**J. Thomas "Tom" Richards** was born in 1852 in Ohio; his parents had emigrated from Wales. His father was a weaver. Tom worked in mines in Bonnersly, Ohio, from age twelve. He preferred to work rather than attend school. After relocating to Erie in 1876, he worked in the mines for a while, but in 1888, he opened a general store that supplied lumber, groceries and farm machinery. In 1877, he married Elizabeth "Eliza" Thomas from Illinois. They had six children. Later, he married Belle Neff (1872–1955) from Kentucky. She brought her daughter, Ruth, with her. In 1898, Tom became postmaster and located the town's mailboxes behind the counter of his store, which he operated until 1906. He also managed one of Erie's early newspapers, *Erie Review*. He died in 1924 and is buried in Erie's cemetery. Tom's sister, Charlotte "Lottie" Richards, moved to Erie at the age of fourteen to be closer to her brother; she eventually married James L. Wilson.

**John A. Rowe** was present at Erie's incorporation in Greeley in 1874 and served on Erie's first board of trustees.

**Jim Sharratt** was killed at the Washington Mine when the roof of the mine caved in under him.

**Jean Slater** was the daughter of James L. and Charlotte (Richards) Wilson. She and her siblings—Fred, Stella and Francis—were born in Erie. Jean worked in the Wilson-Larson General Store for many years. She married Harold Larson in 1921. She remarried to Will Slater of Longmont around 1951.

**Ray Smith** co-organized sixty miners and formed Local 711 of the Knights of Labor, the first national union in Colorado.

**Walter W. Smith** was born in 1853 in Ohio. He moved his drugstore into 502 Briggs Street in 1893.

**Sam Southard** owned a livery stable; he also served as town clerk and assessor.

**Arthur Spendlow** was born on January 14, 1891, and started mining at age thirteen. His father was killed in an Erie mine in 1891. At twenty-four, he was tried for attempted murder following the Hecla Mine incident

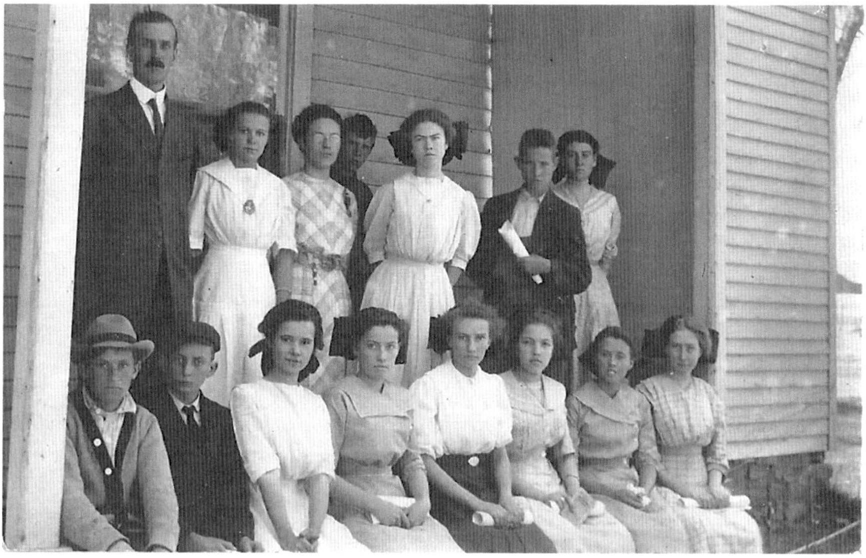

Jean Wilson (front row, fourth from left) poses with her eleventh grade class in 1910. She graduated from the State Teachers College of Colorado in 1913. *Sue Larson collection.*

in which one of the fire bosses was shot in the hand. Actually, eleven men were indicted, but only four were to be found when the trial started: Arthur Spendlow, William Knowles, Dan Griffith and Gus Brack. Arthur was a bachelor and the cousin of Daisy Bracegirdle. He and Brack were acquitted, while Knowles and Griffith were convicted. He died on May 26, 1962, and is buried in Erie's Mount Pleasant Cemetery.

**Robert Summerfield** moved to Erie in 1918 with his wife, Cora. He worked as a foreman for the railroad. Robert and Cora were active members of various lodges in Erie. They moved to Erie so that Dr. Bixler could attend to their daughter.

**T.J. Turpin** moved to the United States from Scotland. He was crushed at the Puritan Mine in 1930.

For information on **Reverend Richard Jeptha van Valkenburg**, see Chapter 5: First Churches.

**Enoch Vaughn** was a coal miner from England. In 1892, he erected a one-story Italianate brick house at 405 Holbrook Street. His wife, Mary, died in 1898. Enoch eventually moved to Como in Park County with his daughter, Minnie. Thomas Hunter later owned his house.

**Henry Adolph Waenekae** came to the United States from Germany and eventually settled in Lafayette. Several of his family members did, however, live in Erie. The name underwent spelling changes, including Waeneke and Waneka. Today, family members use the Waneka spelling. Well-known relatives of the Waneka family in Erie include last names such as Rodick, Jordan, Mosnik, Gorce and McAfee.

**Charles Wallis** helped to plat the Canfield area.

**William Walters** was born in England. He worked for the Boulder Valley Coal Company at the Centennial Mine, where he was crushed and killed between two coal cars in 1931.

**John C. Wells** was the secretary of the Boulder Valley Coal Company when Ransom Balcolm was president. The founders of the company were instrumental in the platting of Erie. Wells Street is named after John.

**Henry T. West** was the proprietor of the Rob Roy Mine in Canfield in 1879.

**Joseph J. Wharton** was present at Erie's incorporation in Greeley in 1874 and served on Erie's first board of trustees. He was married to Caroline Wharton. Joseph lived from 1839 until 1903 and is buried in Erie's Mount Pleasant Cemetery. Caroline died on September 22, 1897, and is also buried in Mount Pleasant Cemetery.

**William "Billy" Whiles** came to Erie from England. His uncle, James Pallott, operated the Garfield Mine. Billy worked ten-hour days in the mine, and he spent his evenings studying stationary engineering through a correspondence course. After he graduated, he became a hoist engineer for the Garfield Mine. He eventually owned the mine, along with William H. Drinkwater, another young Englishman. In 1903, Whiles left mining to work for the Erie Bank as a teller; he eventually became bank president. He gained the confidence of residents when he foiled a bank robbery. In 1908, Billy joined others to file for incorporation of the town of Firestone. Whiles also petitioned for the establishment of Dacono. He and his brothers sunk the Warwick Mine in east Evanston (Frederick) in 1906. He started the First Bank of Frederick.

**Thomas William Wiggett** came to Erie in 1879 aboard a boxcar on the Union Pacific Railroad. He and his wife, Maria, moved all of their belongings from the boxcar into a cave near Coal Creek, where they lived until they could afford proper housing. Eventually, they moved into an adobe house on Briggs Street. After conditions at the mine discouraged Thomas from being a miner, Thomas and Maria accepted work with the Union Pacific Railroad.

**B.M. Williams** helped to plat the Canfield area.

**John T. Williams** was born in Wales in 1815. He and his wife, Julie, came to the United States and eventually to Erie. They had a son, John, who was born in Colorado. John T. operated a general/hardware store and pharmacy, and he later partnered in his business with John Probert. He also served as town treasurer. He owned houses at 475 Cheesman and 485 Holbrook; John Probert later purchased the Cheesman Street home. Williams was a town trustee in 1875 and president of the board of trustees in 1875–76. He served as reverend to Erie's Welsh Church in the 1890s. He died on November 14, 1896, and is buried in Mount Pleasant Cemetery.

**Francis A. Wilson** was the son of James L. and Charlotte (Richards) Wilson. He and his siblings—Fred, Jean and Stella—were born in Erie. Francis married Ruth Neff (1893–1976), daughter of Belle Neff Richards. They had three children, Richard, James and Raymond. Francis served as mayor of Erie from 1930 to 1932. He also opened the first gas station in Erie and had an automobile repair business. He and Ruth are buried in Boulder's Mount Pleasant Cemetery.

**Fred R. Wilson** was born in Erie in 1888 to James L. and Charlotte (Richards) Wilson. He had three siblings, Stella, Jean and Francis. Fred learned to run stores in Hanna, Wyoming, and then returned to Erie in 1924 to run the Wilson-Larson General Store. He and his wife, Rhoda, had two children, Dorothy and Olive. The girls were both born in Hanna around 1915. Fred was active as a town board member, firefighter and charter member of the Lions Club. Fred served as Erie's mayor three separate times: 1932–34, 1948–50 and 1955–57. He also worked with the Boy Scouts and an organization to assist the blind. He was a member of the Erie Methodist Church. Fred co-owned the Wilson-Larson store with his brother-in-law, Harry Larson, until the 1950s. He died in 1973, and he and Rhoda are buried in Boulder's Mountain View Cemetery.

**James L. Wilson** moved to Erie from Rochester, New York, where he was born in March 1858 to Alexander (born in 1826) and Celia Lyon Wilson. His grandmother, Janet Clark Wilson, arrived in Rochester in 1833 from Dalkeith, Scotland. A brick maker and builder, James constructed many of the houses in Erie. He also built the Methodist church. He formed the J.L. Wilson Fire Hose Company. He was a member of the Knights of Pythias. On the 1907 graduation commencement announcement from Erie High School, Wilson is listed as the president of the board of education. James was mayor of Erie from 1888 to 1891, 1893 to 1894 and 1921 to 1922. He was also the justice of the peace. He married Charlotte "Lottie" Richards in 1887. She was the sister of J. Thomas Richards; she moved to Colorado from Ohio at the age of fourteen with her sister to be near her brother. James and Charlotte had four children: Frederick Richards (born

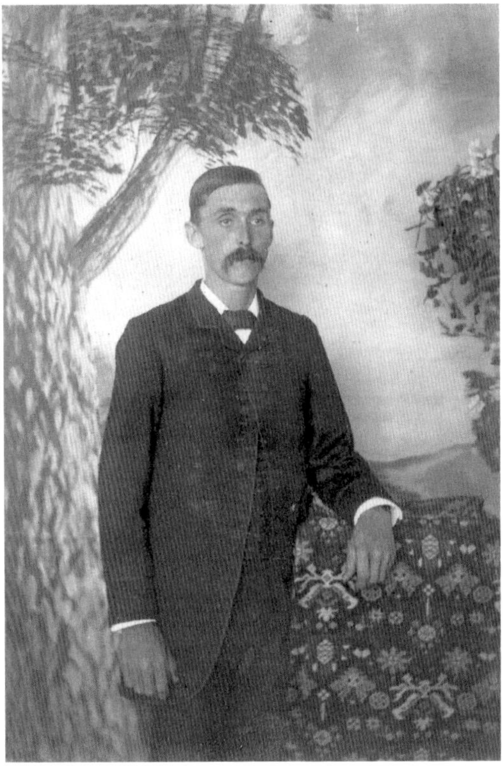

James Lyon Wilson. *Sue Larson collection.*

in March 1888), Francis Alexander Hope (born in March 1892), Jean Evelyn (born on February 23, 1894) and Stella Charlotte (born in November 1896). All were born in Erie. In 1887, James built the family home at 724 Holbrook. In 1938, he died from injuries suffered from being struck by an automobile; he was eighty-one years old.

**Joseph O.V. Wise**, son of Oliver and Adeline Wise, was born around 1853 in Wisconsin. He had one brother, William O., and one sister, Ada. He married Sarah Beasley, born in 1857; they had three children. He donated the land for the Canfield School; he was treasurer of the school district. He planned and helped build some of the local irrigation systems. He and others platted the Canfield area. He was the proprietor of the Standard Coal Company.

**Oliver Ellsworth Wise** left Dodgeville, Wisconsin, to settle in Colorado in 1869. He arrived with his wife, Adeline, and their three children, William O., J.O.V. and Ada. Receiving his right to homestead in 1870, he was one of the original homesteaders in Canfield, about a mile west of Erie. Their homestead was near several mines, and they had the opportunity to help develop a community to serve the families in the area. Oliver operated a blacksmith and general store in the Canfield area. Jasper Road, where the Wise History Homestead is located, used to be called Wise Road. Oliver provided goats for the Greek community in Erie.

**William O. Wise** was born in Dodgeville, Wisconsin, in either 1848 or 1850. After recovering from childhood illness, he originally worked as a farm hand. He became a decent poet, writer and newspaperman, taking time off to form the *Erie-Canfield Independent* newspaper. Later, he served in the Colorado State Legislature. With his brother, J.O.V., he was a joint owner and operator of the Standard Coal Company. He was superintendent and part owner of the Star Mine, along with co-owners Joseph C. Dresser and Benjamin M. Williams. He and J.O.V. co-operated the general store and post office in Canfield. William was postmaster for ten years. He was a stockholder in and director of one of the first flour mills in the area. William successfully petitioned the governor in 1882 to establish a post office in Canfield. He served for fifteen years on the Canfield school board. He was also one of the eleven people named on the November 11, 1878 incorporation papers for the "Baby Railroad," the narrow gauge train that ran from Erie-Canfield to Longmont. He was married to Margaret Delphia Matthews, and they had three children: Gertrude P. (1874), Grace (1879) and L. Gladyce (1890).

**William T. Wise**, son of J.O.V. and Sarah, operated the feed mill in Canfield. He had suffered from polio in childhood, and he retired early because of his physical disability. His wife was Ella; they are the parents of Dr. Sarah Allene Wise of the Erie Historical Society.

**Mable Young** was Erie's first teacher in the original school at Wells and Briggs Streets.

# 3
# THE FOUNDING OF ERIE

The community in the southwest corner of Weld County was nameless. Many called it "coal park." In 1871, the town was platted under the name "Erie," and on July 24, its post office, one of the first in Weld County, served about fifty people. Richard van Valkenburg was postmaster. The plat showed blocks measuring 320 by 400 feet, with thirty-two lots per block. By 1872, Erie had six hundred residents and had become the third largest town in Weld County.

An adjacent community called Canfield sprang up in 1869, at what today is 119th Street and Jasper Road. Isaac Canfield mined in the area, but he soon abandoned his efforts. A few of the area's best-known mines were located there, and the railroad eventually served the town at the Tabor Stop, named after Horace Tabor of Denver. Today, while Canfield is a part of Erie, some of the land is unincorporated.

On November 16, 1874, the town of Erie was established. Five residents oversaw the initial incorporation of the town and became the first board of trustees: George Meller, Richard van Valkenburg, John T. Williams, Joseph J. Wharton and John A. Rowe. Richard van Valkenburg named the town after Erie, Pennsylvania, a coal mining town near his former residence.

By 1875, businesses lined Briggs and Kattell Streets. Erie differed from other coal mining towns in that some miners actually owned homes and helped to develop the town. Some raised chickens and planted gardens. Nearly everyone lived and traded in Erie, Canfield or Lafayette. The *Boulder County News* referred to Erie as "one of the most lively little towns in this part of the country." It was the only "wet" town in the area. In 1895, people had their choice of eleven saloons on Briggs Street. One estimate is that Erie's population in 1883 was around eight hundred. Perhaps they were including migrants and farmers who lived in the greater area.

As Erie grew, so did the need for rules and a place for those who had difficulty following them. In 1876, a jail was constructed, and a town constable was appointed. One of Erie's earliest constables was B.C. "Bud" Pitchford.

In 1883, Erie annexed its first subdivisions—the West Addition and the East Addition. These were merely expansions of the previously platted town. The East Addition was basically the land between Kattell Street and Coal Creek, and the West Addition was west of Holbrook Street, eventually known as Erie Heights.

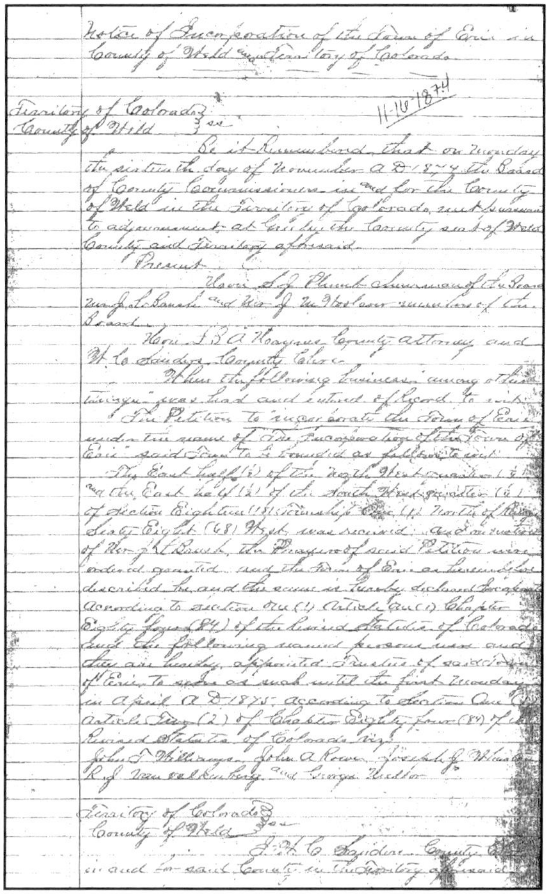
Notice of Incorporation of the Town of Erie in County of Weld and Territory of Colorado.

11-16-1874

Territory of Colorado } County of Weld } ss

Be it Remembered, that on Monday the sixteenth day of November A.D. 1874 the Board of County Commissioners in and for the County of Weld in the Territory of Colorado, met pursuant to adjournment at Greeley the County seat of Weld County and Territory aforesaid.

Present

Erie's Notice of Incorporation in 1874 differs dramatically from such documents today. It was handwritten into a ledger. *Town of Erie.*

Until people could move into houses, many lived in tents, mold huts or dugouts along Coal Creek. As more people with a greater variety of skills moved in, houses appeared in town. Richard van Valkenburg is credited for building Erie's first house. Later, James Wilson built many of Erie's houses. The earliest houses on record were built around 1870.

Mining companies frequently established their own camps and provided housing and stores for the miners. Miners who lived in company housing

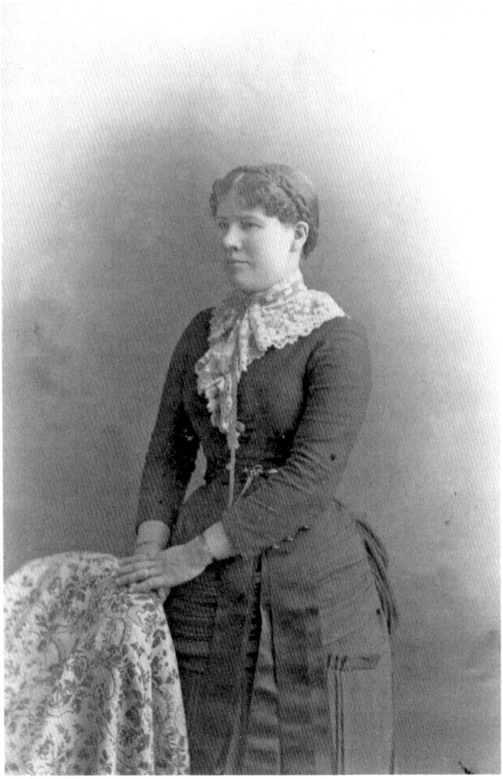

Charlotte "Lottie" Richards Wilson. *Sue Larson collection.*

were usually required to buy their clothing, groceries and hardware from the company stores, using scrip. Prices in company stores were higher than those of the stores in town. Accounts were settled when money was deducted from miners' paychecks.

Erie did not have a natural source of drinking water. People relied on a well dug by the Union Pacific Railroad, originally intended to provide water for the railroad's steam engines. Later, the community purchased water from an agricultural reservoir. Because Erie had no water treatment facility, the usefulness of the water was limited. Water from the local creeks was unfit for drinking.

In 1885, Erie passed a bond measure that allowed drilling for three artesian wells—one at Holbrook and Cheesman Streets, one at Moffat and Pierce Streets and one near Briggs and Perry Streets. The water was accessible from water hydrants at each location. Although the wells were not used after a water treatment plant was eventually located in the town in 1940, some homes along Briggs Street still use well water today.

Early in Erie's development, people formed a number of social organizations, including the Masons, the IOOF, the Good Templars, the Knights of Pythias, the Pythian Sisters, the Maccabees, the Rebekah Lodge and more. The IOOF Lodge 46 received its charter in 1882. William Fletcher was its first Grand Noble. The Garfield Lodge Number 50 was Erie's Masonic branch. Its charter members included Reverend Richard van Valkenburg, W.R. Howell, John Marshall, Joseph Morgan and George Grenfell.

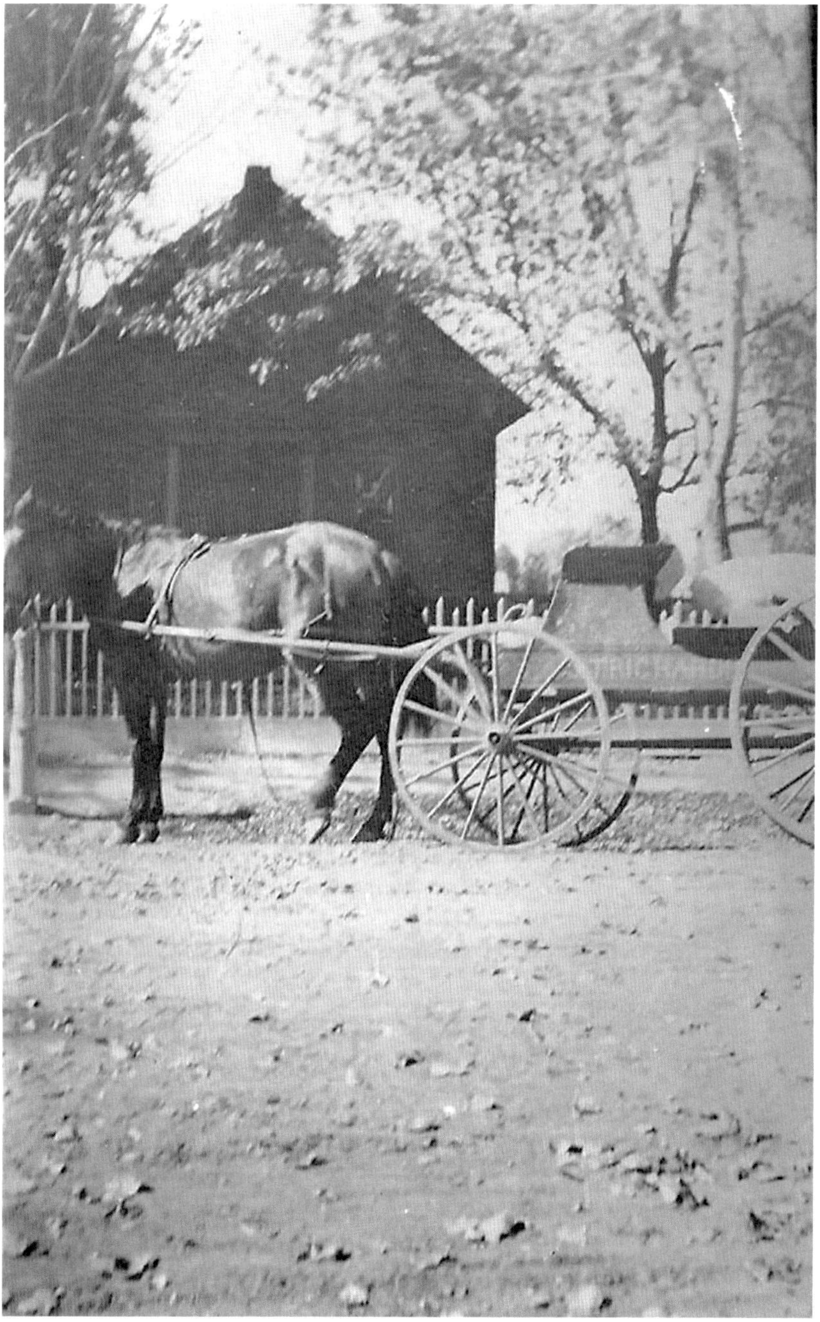

Babe was the horse Tom Richards used to pull his delivery wagon. Here, Babe and the wagon are stopped at James L. Wilson's home on Holbrook Street. *Ray Wilson collection.*

The Ku Klux Klan was also active in Erie. In the 1890s, the American Protective Association attracted many Coloradoans, and because they had similar goals, they were prime candidates for the Klan. If one was white, male, Protestant and native born, he was eligible for membership. Around 1921, the Klan was visible in Denver, and by 1922, a Klan parade was held in Boulder. Some people did not try to hide their affiliation with the Klan, including Colorado's governor Clarence Morley, some Colorado senators and Denver's mayor Ben Stapleton. Other members were Denver's police chief Bill Candish and Colorado's lieutenant governor, secretary of state, attorney general and Supreme Court justice. Nationally, Presidents Warren Harding and Harry S. Truman and countless other prominent figures were believed to be members. Some newspapers and businesses also showed support for the Klan, likely to avoid trouble. Denver, Longmont and Boulder had the largest Klaverns in Colorado.

In 1926, Klan members marched into the Erie Methodist Church during a service. Reverend W.J. Keeser asked them to leave when they refused to remove their hoods. He also requested they give a contribution on their way out of the building. The Klan went immediately to a hill east of town and burned a cross. Nothing further developed.

Allegedly, most of Erie's Klan members were miners who feared younger and outside miners taking their jobs, particularly during times of unrest at the mines. They weren't the stereotypical Klavern that attacked Jews, Catholics, African Americans, Hispanics and Eastern Europeans. A women's auxiliary of the Klan was known for performing charitable and benevolent acts.

Life in Erie was not all work. Children played along the creeks, trying to catch muskrats, skunks and snakes; picking flowers; playing in the caves north and east of town; and carving their initials in the sandstone. In the summertime, they played street games like "kick the can" or "run sheep run." They played baseball with broomsticks and wadded-up socks.

Adults formed baseball teams. Erie, the Puritan Mine and Canfield each had their own teams with uniforms. The best place to play was on the flatlands along Coal Creek. Families packed picnic lunches and came to watch. Brass bands would often play. These games became major social events.

Dancing was also popular. In 1904, people attended the Erie Leap Year Club dance and enjoyed various dances they brought along with them from Europe. Once the annex next to the Methodist church was converted into a community center in the 1920s, dances were held there. Square dancing was done to live music at Jimmy Smith's barn, the Jarosa barn near Canfield and

Momford's barn east of Lafayette. People from all over Boulder and Weld Counties would attend and stay until early morning.

Erie residents formed brass bands that played at picnics, dances, baseball games and other events. They frequently accompanied striking miners as they demonstrated at the mines. Miners liked to play in the bands, and some even said that blowing their horns made them feel as if they were ridding their lungs of coal dust. The Erie Brass Band was formed in 1875, with some of the founding members listed as Harry Shorthouse, Thomas R. McClure, James Beveridge, David Davis, D.B. Grubbs, S.J. Foote, Joel Palmer and E. Lafferty.

Movies were shown in the community center and at the Opera House in the 1920s. For just ten cents, children could watch cowboy and Tarzan movies. As a youth, DeWitt Brennan sold popcorn there. Pauline Phillips and Lucille Carter played piano interludes while film reels were being changed.

Mike Brennan supported his family in the summertime by providing bingo games at fairs in the nearby communities. Bill and Singer Davis operated a bicycle and horse track south of Erie. Erie prohibited any kind of racing within the town limits. People bet on the outcomes. A roller-skating rink was built on Briggs Street, where people could skate for two hours for twenty-five cents. The rink was open until midnight, and live music was often provided. James Waneka remembered that during the winter, the town would flood the high school tennis court so residents could enjoy ice skating.

People enjoyed high school athletic and theater performances. Erie's baseball and basketball teams were very competitive. The girls' basketball team won the state of Colorado championship in 1929. The boys' basketball team remained undefeated in league play in 1935, only to be defeated in the state tournament.

Erie's first newspaper was the *Erie-Canfield Independent*, started by Charles D. Bell and published weekly. A few years later, the *Erie Independent* became a second newspaper, owned by the Jones brothers. By 1896, neither newspaper was in circulation. Three other papers had short lives in Erie. The *Erie Review* was started by F.H. Davis and Jack O'Dell and served Erie from 1899 to 1912. The *Erie News* was started by Clyde L. Stanley and ran from 1902 to 1906. The *Western Mirror* was the product of John Hoffman, and it served Erie from 1900 to 1907. In 1907, the *Erie Herald* became Erie's primary newspaper, and it served Erie until 1948. It was started by Walter McAnally and later sold to a Mr. Olin and then to Roy W. Hays. While visiting one of the older Erie homes, the author was shown a 1935 copy of the *Erion*, Erie High School's paper.

Erie celebrated special holidays that were meaningful only to the local communities. April 1 was John Mitchell Day, when people honored the president of the United Mine Workers. Mitchell was responsible for negotiating an eight-hour workday for miners. School was canceled, and residents attended street fairs and parades. This holiday was observed as far away as Denver, and it continued to be observed until the 1960s. Labor Day was also important to the union town of Erie. People came from all over the area to observe the holiday, partially because Erie served alcohol.

Another festive occasion that has returned to Erie after not being celebrated for many years is Biscuit Day. Every October, the local baker, Chris Miller, gave away fresh biscuits with apple butter. People purchased bowls of stew to accompany the biscuits. Ladies ate for free. On this day only, people were allowed to race their horses down Briggs Street. This holiday was established to celebrate miners going back to work in the mines, as the mines closed during the summer, and people left for summer work in the crop fields or mountain mines. Today, Biscuit Day is in September, with local restaurants providing stew and competing for the reputation for having the best-tasting stew.

On July 4, Erie held a parade in the morning, led by Erie's brass band. The parade ended at the ballpark. Children bought fireworks at Elzi's Drugstore,

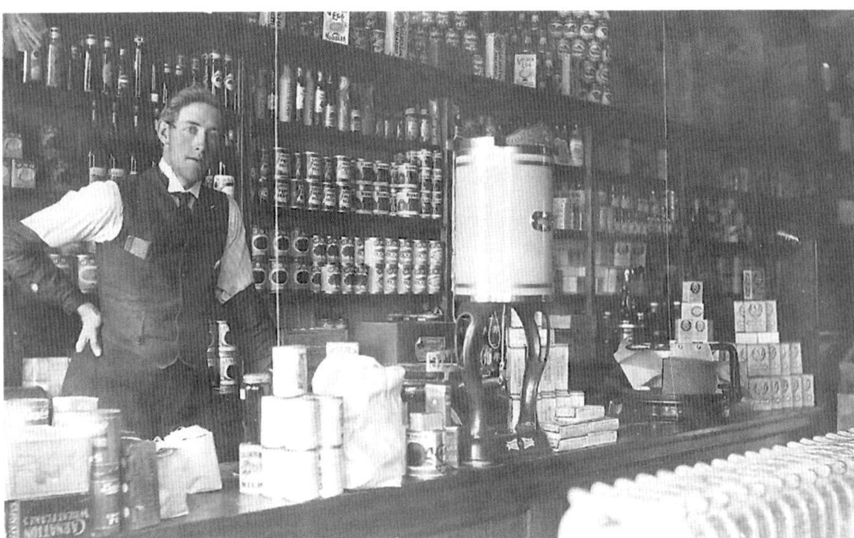

A young Charles Elzi poses behind the counter of his drugstore at 502 Briggs Street. *Ray Wilson collection.*

and the Brennan boys brought fireworks, compliments of their employer, the *Denver Post*. Caution was observed, as Mike Brennan lit the fireworks in a rain trough.

During the Great Depression, the United States government created numerous agencies to help people earn a living and preserve their dignity. Because Erie's mines had to shut down, the Works Progress Administration funded projects that Erie's miners and other residents were able to complete. Some of these projects included improving streets and buildings, developing new tennis courts north of Erie High School and constructing dikes along the banks of Coal Creek to protect the town from flooding. The government also established a food program. The Depression and summer mine closures forced miners and their wives to work in the fields, picking fruits and vegetables. Men often hunted rabbits and sold them in the streets. Soup lines were common. Local merchants cut prices and extended credit at their stores. Even during the Depression, residents enjoyed dances, movies, baseball games, band concerts, radio programs and competitions from fire hose teams from each mining camp town. Train rides were fewer, and automobiles became scarce. To make matters worse, in 1933, a grasshopper plague hit eastern Colorado.

Erie children played in dugouts or caves near town, which, they were told, had existed since the days of the Arapaho and Cheyenne. They had to be careful, however, because displaced families occupied some of the caves. Everyone watched for the bands of gypsies who would wander through the area.

Harold Conroe is a longtime Erie resident. He served as mayor in the 1960s, as well as a teacher, counselor and principal of Erie High School. Even that long after the depression, he reports Erie as being "poor, poor, poor." Marlene Mosnik Waneka was born in Erie and remembers being very poor and living in a house with no running water.

Erie's first cemetery is believed to have been near the Welsh Presbyterian Church on Holbrook Street. Some people argue that the cemetery was endangered because it was on a flood plain. There is no evidence the floodwaters ever reached that far west into Erie. The cemetery was eventually relocated east of town above Coal Creek and north of Erie Parkway on Weld County Road 3. People question whether all of the actual graves were relocated from town. The new site was called Mount Pleasant Cemetery (or the Erie Cemetery). It overlooks downtown Erie and provides a beautiful view of the Front Range and prairie leading up to it. The land was originally provided by the Union Pacific Railroad to bury one of its depot agents in 1882.

Mount Pleasant is a natural prairie cemetery, one of the relatively few remaining in the United States. The grounds have not been covered with grass nor have they been irrigated. Moisture is provided by nature or plot owners, who care for some graves. Tulip, iris and other bulbs provide floral beauty in the springtime. The Town of Erie's Parks Department provides monthly maintenance.

In the earliest days of Erie, the deceased were typically transported by horse and wagon to the cemetery. Many of those buried were victims of influenza epidemics in 1910 and 1918. Others fell to the diphtheria epidemic of 1930. Some graves belong to miners who died in mining accidents. As of 1967, someone had counted 274 unmarked graves. Because the town had poor, incomplete or non-existent records in its early days, many other graves may be unmarked. People were often buried in sections based on their ethnicity. Dozens of grave sites exist on the hillside outside the marked perimeter of the graveyard. A tour of the cemetery is a history lesson, chronicling the deaths of miners and other residents of Erie's earliest days.

For more complete information on the symbolism on the headstones, contact the Erie town clerk. Rather than listing here all of the nearly one thousand people interred at the cemetery, the author encourages you to Google "Erie Colorado Mount Pleasant Cemetery." The site even provides links to pictures of some of the deceased and their gravestones.

4

# HISTORIC BUILDINGS

If people tour Erie today, they will notice diverse architectural styles among Erie's houses and buildings, from Victorian and Queen Anne designs of Europe to false-front Western-style stores and geodesic domes.

By 1873, about one hundred homes existed in the area. The 25- by 150-foot lots were available for sixty dollars each. Houses were affordable to attract and keep miners in the area and to discourage them from striking. James L. Wilson built many of the houses. Houses drew water from artesian wells; outhouses were built in the backyards. One Erie pioneer recalled how everything in the house froze some winter nights—even shoes stuck to the floor.

Many of the following buildings were researched in 2010 to determine their historical significance. They are only a sampling of Erie's treasure of old dwellings and commercial buildings. Literature is available throughout the town for those wishing to take a self-guided tour. To encourage people to visit the sites, only a few photos are included here.

The **Wise Homestead Museum, 11611 Jasper Road.** Oliver and Adeline Wise built this home in 1870 as a permanent building on their homestead. It is a Western Victorian–style building, one of the oldest frame houses in Boulder County. The home is now a museum, which has been given to the Town of Erie. It has two bedrooms upstairs and one downstairs. The original downstairs also had a kitchen. A large livingroom, a dining room and two other small rooms were added, as well as a bathroom with plumbing. The original wallpaper contained mica, which helped to light the room better as it reflected candle and lantern light.

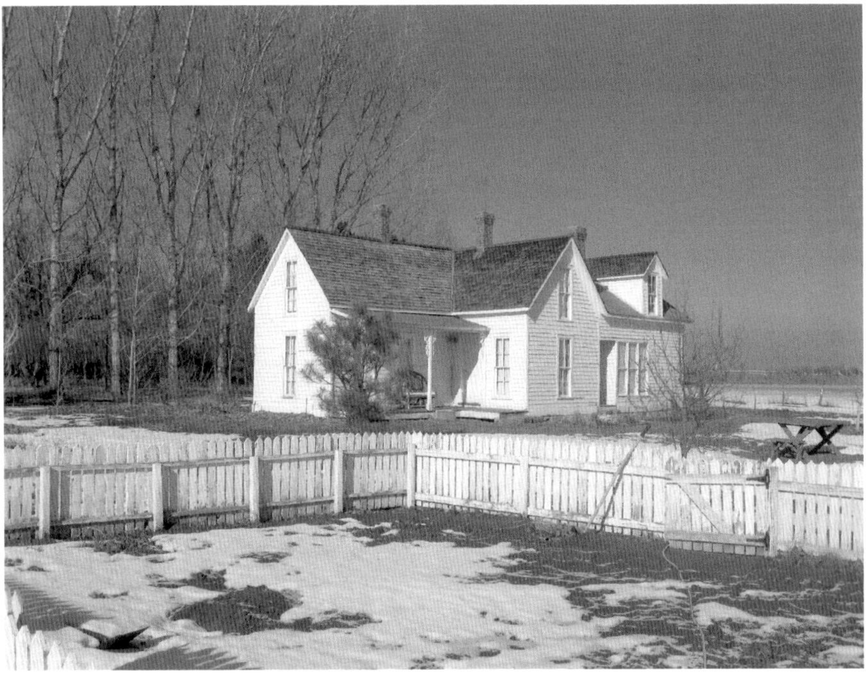

The Wise Homestead was one of Canfield's earliest structures. The Erie Historical Society recently donated the museum to the Town of Erie. *Author's collection.*

**Powell House/Egnew Hotel, 370 Briggs Street.** This hotel is estimated to have been constructed sometime between 1890 and 1892. It served as a boardinghouse and hotel until the late twentieth century. At one time, it was called the Erie Hotel or Erie House. It was the second hotel in Erie and is still at its original location.

Alexander Morrison, a Scottish immigrant who worked in the mines, provided the land for this structure. He turned the property over to Susan Thompson of Greeley, and she sold it to Alice Powell, the hotel's original owner and daughter of Richard van Valkenburg. Alice was married to Joseph R. Powell, who, with William Nicholson and George Grenfel, owned the Long's Peak Coal Company.

The hotel had fourteen rooms. The downstairs had a lobby and gambling room in the front, a dining room behind that and a kitchen at the rear of the building. Upstairs were the sleeping quarters.

The hotel changed ownership over twenty times until it was known in the 1990s as the Ash Pit Flophouse and Grill. The owner, Hans Bowker, started taking in former prison inmates; the last shootout and murder in the hotel under his ownership occurred when a local prostitute was refused payment by her client. In 1994, Hans was the campaign manager for Jake the Booner, a dog whose name ended up on the voting ballot for town mayor.

**Angove/Wood House, 455 Briggs Street.** This house is thought to have been constructed as a rental; it is an example of the typical architecture in the early 1900s in Erie as a hipped-roof box house. It is also associated with the Zimmerman and Hauck families. The title to this house changed many times until 1978. William Angove was from England; he operated the Angove and Hershey livery and feed store and a grocery store. He was a teamster, and he served as town treasurer. Today, this home has an enclosed front porch and a wheelchair ramp.

**Multipurpose building, 485 Briggs Street.** People tell different stories about this building, and little is known for sure about it. It was built in 1888. It served as a bank and Erie's post office. Bob Swenson, formerly of the Denver Broncos, owns it.

**Independent Order of Odd Fellows (IOOF) Hall/Lockwood Trading Company/State Mercantile Company/Hunter Mercantile Company/Wilson-Larson General Store, 500 Briggs Street.** On the northwest corner of Briggs and Moffat Streets sits this two-story, orange brick building, originally built in 1889 as a meeting place for Lodge 40 of the IOOF upstairs, with a business facility downstairs. It represents nineteenth-century commercial design. It was once known as the Thomas Hunter Building, with a number of stores providing groceries and general merchandise for Erie's residents. In 1892, it housed the Lockwood Trading Company.

John T. Williams owned the original property. He sold it to the IOOF in 1885. The title on this structure transferred several times. For a few years, it was vacant. The Erie Lions and the IOOF jointly tried to renovate the building. It is thought that maybe the building at 502 Briggs was built to expand this building. This building is now the Davies Building and recently has been home to the Kabbu Bar, the Whimsy Paint and Sip Art Studio and The Old Mine.

**Winslow and Smith/W.W. Smith/Elzi Drugstore, 502 Briggs Street.** This long, rectangular, one-story brick building was originally built to be a commercial trade/specialty store around 1892. It served as a drugstore in Erie. It changed owners several times; most notable were the Colorado Coal Company, IOOF, Winslow Smith, James Brennan, Charles Elzi and the Erie Lions Club.

Charles Elzi came to Erie in 1912 and worked in Walter W. Smith's drugstore. After Walter's death, Elzi operated his own drugstore until just before his death in 1972. In 2010, it was the location of Zupa European Café, owned by Katherine R. and Piotr H. Wolak. They purchased the building in 2005. Since, it has housed a consignment store called Huckleberry, the Flower Nook and Sweets, an ice cream and coffee house.

**Erie Bank, 512 Briggs Street.** Erie Bank was Erie's first bank. It is now the location of the First Bank of Longmont–Erie Branch. In 1932, when President Franklin D. Roosevelt ordered a national bank holiday, Erie Bank continued to operate because it had sufficient assets. Many banks were unable to reopen following the holiday. The inside of the original bank has been remodeled many times. Erie Bank was robbed in 1956 and 1959. Nettie Waneka worked at the bank in 1959 and is believed to have scared off the robbers when she arrived at work before they could take much money.

**Morgan's Bar, 545 Briggs Street.** This false-front commercial building was erected between 1908 and 1948 as a restaurant and bar. Probably one reason for the lack of certainty about when this establishment was built is that three similar buildings were originally in the same block, and records were vague. This building was listed as 549 Briggs in the 1964 and 1969 directories.

It is possible that cigar maker Henry Corcilius used this as his store. One of the other buildings with the same architectural style may have been the location of J.T. Richards's first store.

More than twenty owners are listed in the title history of this property. Some of the merchants were John Marshall, a livery stable owner; Charles Evans, a dry goods store owner; Walter Vinovich, a bar owner; and Sandra Lou Louis, a health food store owner. It has also served as a video store. Most notably, though, it was Joe Morgan's bar from 1928 until World War II. He installed a nickelodeon, offered candy and sodas and permitted school children to go there for thirty minutes after school before the miners arrived after work.

**D&F Groceries, 600 Briggs Street.** This building burned down, but it used to served as a mercantile. In the 1950s, it was D&F Groceries, operated by Donald Zaruba and Fred Hasenack. Fred was Donald's stepfather and lived at 534 Holbrook. The family suffered a major tragedy in 1957 when Irene Hasenack and Cecilia Jordan (mother of Lorene Zaruba) were killed in an automobile accident on U.S. 287 in Lafayette. Donald Zaruba, son of the store co-owner, was seriously injured in the accident. Lorene was also mother to Ronald and Debbie (Langerak).

Since the building burned down, the location has been the home to a church, firehouse, Erie's community center and the Arts Coalition of Erie. At the time of this writing, it is the site of Echo Brewery.

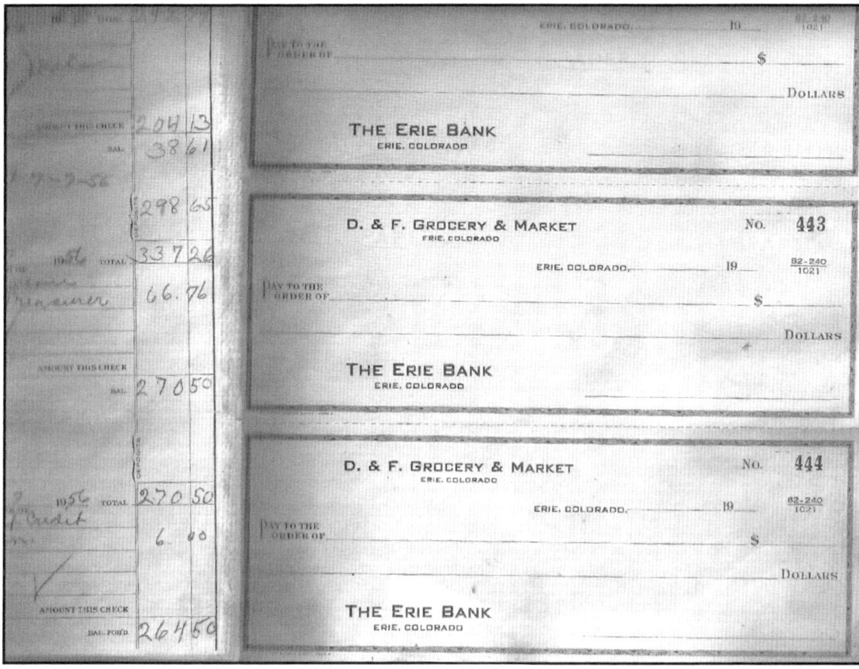


Donald Zaruba and Fred Hasenack used this original checkbook at Erie's D&F Grocery. *Debbie Langerak collection.*

**Williams/Probert House, 475 Cheesman Street.** This late Victorian house was built around 1884 and is associated with John T. Williams and John and Sarah Probert.

George and Lucretia Winslow purchased the house in 1893. George was co-owner of Winslow-Smith, a provider of farm implements. He soon sold it to Gwennie Williams, who later sold it to Sarah Probert. Sarah sold it to her daughters; it stayed in the Probert family until 1968.

**Walker House, 574 High Street.** This late Victorian house was built around 1894 in Erie's new subdivision called Erie Heights. Its original owners were Franklin D. and Marvel May Zook Walker. Frank was a teacher, the mayor of Erie and a carpenter. Marvel May was a teacher and served as the secretary/treasurer of the Colorado Education Association. They lived in the house until approximately 1947. Franklin D. Walker and Cynthia Garnier gained the title in 1980.

**Nicholson House, 604 High Street.** This late Victorian/Queen Anne–style house was built in 1895 by William Nicholson, who came from England. The *Greeley Tribune* called it "the finest residence in Erie." William began working in mines at the age of eight. At thirty-one, he moved to Erie and became a mine superintendent and later founded Long's Peak Coal Company. He eventually sold his interest in the company to the Rocky Mountain Coal Company.

William founded and served as president of the Erie Bank. He was Erie's mayor for several terms. He was a member of the Masons and the Elks and a trustee of the Erie Methodist Church.

**Vaughn/Hunter House, 405 Holbrook Street.** This late Victorian/Italianate–style house was built in 1892 by Enoch T. Vaughn, a coal miner. Enoch lived there with his wife and seven children until 1903, when they sold it to Hugh Graham. Graham immediately sold it to Rachel Hunter. She held title until 1921, when she sold it to Thomas B. Hunter, associated with the Hunter Mercantile Store. The house was a square, wooden structure with a parlor, dining room, bedroom and kitchen. A chimney at the peak of the roof vented a potbelly stove.

**Williams/Bixler House, 485 Holbrook Street.** This late Victorian house is thought to have been built around 1884 and first lived in by the John T. Williams family. Williams sold the house to Dora Williams Morgan and Thomas L. Morgan in 1891. Their daughter, Retta Wooley, bought the house in 1905.

Clarence Bixler moved to Erie in 1907 to open up a medical practice; he eventually bought this house from Wooley around 1909. He had moved to Colorado from Ohio to pursue his medical degree from the University of Colorado. Early in his career, he may have lived in the back of the house and used the front for his practice. He was a very popular doctor, and soon he needed to expand his office to a building on Wells Street. He sold the house in 1937, and it has changed hands many times since. Eventually, he moved his practice to Longmont.

**Lawley/Milanovich/Johnson House, 575 Holbrook Street.** This late Victorian house is estimated to have been built in the 1880s. Some of its unique characteristics are its gabled roof, decorative verge boards, projecting porch with spindles and tall windows. Robert and Martha Lawley purchased it in 1884. Arriving in 1871, Robert was an early resident and trustee of Erie. Robert came from England, and Martha came from Wales. It is believed they took in boarders. Sam Milanovich bought the house in 1938 from Martha Gossier. He lived there for twenty-four years.

**Richards/Wilson House, 664 Holbrook Street.** This house is estimated to have been built in 1885. Historians believe Thomas Richards may have built it. His second wife, Belle Neff Richards, was from Kentucky.

Belle received the title to the house in 1924. She sold it to her daughter, Ruth Neff Wilson, in 1931. Ruth married Francis A. Wilson, one of Erie's mayors. The house was in the Wilson family until 1977.

**Morrison/Charlesworth/Woods House, 675 Holbrook Street.** This late Victorian/Queen Anne house was built around 1897 and was lived in by families involved in coal mining and mine ownership. The property was owned by Elisha Atkins, the Union Coal Company and then Union Pacific Railroad. Once the house was built, the first owner was George Morrison and his wife, Rachel. George was a Scottish immigrant, who rose to become the president of the Clayton Coal Company.

The house was sold to Netta Wooley in 1913 and to Thomas Hunter in 1918. In 1919, it was sold to James Charlesworth, who came to Erie from England to be a coal miner. Later, he opened a confectionary and restaurant. He served in the Colorado State Legislature from 1937 to 1938. The house stayed in the Charlesworth family until 1954, when it was sold to Harry F. and Crista M. Woods. It has been bought and sold

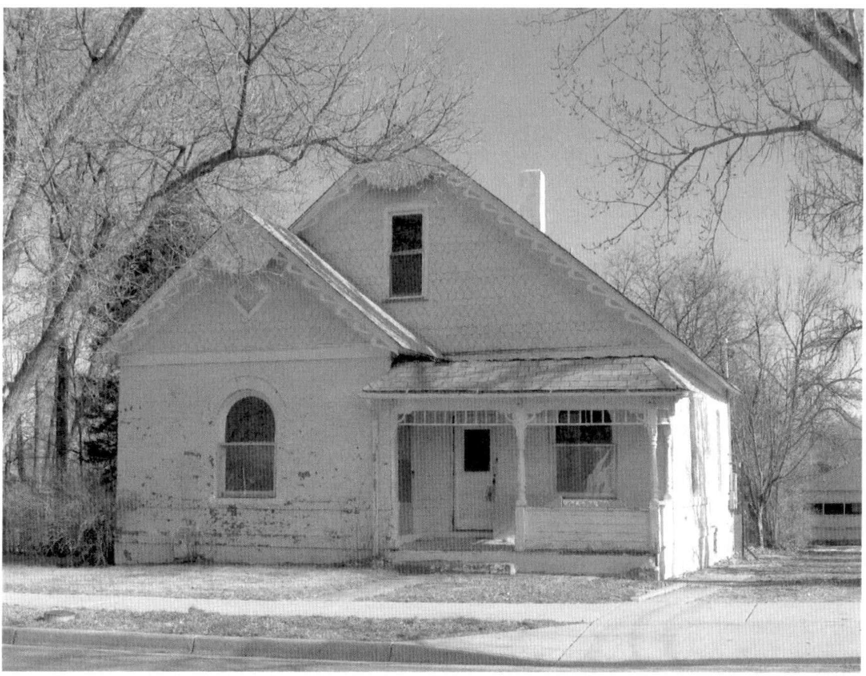

This was one of the many homes in Erie built by James L. Wilson. *Penny Stull collection.*

at least three times since then. In 2013, it was demolished to make room for the expansion of the town hall, perhaps as a parking lot. Members of the Historic Preservation Advisory Board filed injunctions and restraining orders against the destruction, even suggesting that the house be moved to another location.

**Phennah/Brennan House, 684 Holbrook Street.** This property was built around 1889 and was originally owned by Elisha Atkins and the Union Coal Company. In 1891, Martha E. Phennah bought it. She operated a millinery. Mattie's husband, Peter, was an engineer and census taker. They sold the house to Michael and Harriet Brennan in 1917. Michael operated the Erie Opera House, a saloon on Briggs Street and a pool hall. He was Erie's mayor from 1922 to 1923 and served as postmaster in 1941. He was also a justice of the peace. The house has changed ownership at least three times since then.

**Wilson House, 724 Holbrook Street.** Built by James L. Wilson in 1887, this late Victorian house served as his home. Wilson came to Erie from New York and was an early carpenter and brick maker in Erie. He built many of the town's houses as well as the Methodist Episcopal church. Historians regard the porthole-like window on the front of the house as unique and something that may eventually qualify the house to be in a historical register.

**Grimson House/Teachers Boardinghouse, 745 Holbrook Street.** This Victorian house is believed to have been built in 1898. The current owners, George and Debbie Bachelder, possess many of the house's original documents, photos and furnishings. Although the house has undergone major renovations and additions, the original rooms have been maintained and restored. Some of the early families to live there include

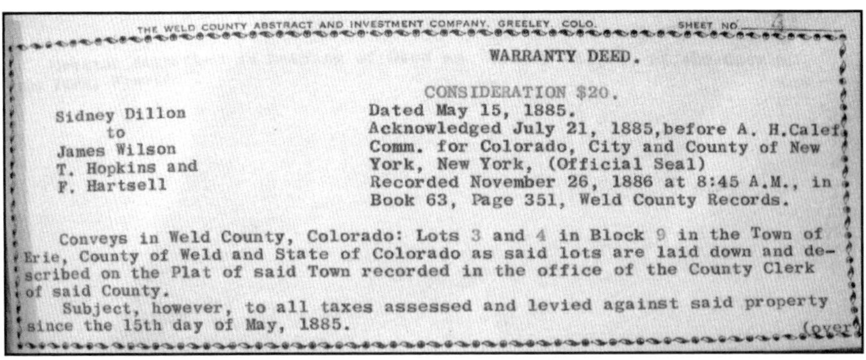
THE WELD COUNTY ABSTRACT AND INVESTMENT COMPANY, GREELEY, COLO. SHEET NO.

WARRANTY DEED.

CONSIDERATION $20.

Sidney Dillon
to
James Wilson
T. Hopkins and
F. Hartsell

Dated May 15, 1885.
Acknowledged July 21, 1885, before A. H. Calef, Comm. for Colorado, City and County of New York, New York, (Official Seal)
Recorded November 26, 1886 at 8:45 A.M., in Book 63, Page 351, Weld County Records.

Conveys in Weld County, Colorado: Lots 3 and 4 in Block 9 in the Town of Erie, County of Weld and State of Colorado as said lots are laid down and described on the Plat of said Town recorded in the office of the County Clerk of said County.

Subject, however, to all taxes assessed and levied against said property since the 15th day of May, 1885.

(over)

This Weld County abstract shows that land was transferred from Sidney Dillon to James Wilson and others. Dillon was a New York attorney for the Union Pacific Railroad. Sidney, Nebraska, is named for him. *Karen Jones collection.*

Brack, Grimson and Gregg. It was listed in a Denver newspaper as having architectural significance around 1902.

As the Lincoln School grew and needed living quarters for its teaching staff, it reached an agreement with the owners of the house to add on a section with several small rooms. A doorbell system still exists today that was used when callers wanted to visit individual roomers. A light and bell system was used so teachers could order their meals from the kitchen.

**Morgan/Carter House, 704 Main Street.** This late Victorian/Queen Anne house was built in 1892. John S. Morgan, from Pennsylvania, was the first owner. His parents were from Wales. He was Erie's postmaster, a town trustee and a merchant. His store was located below the IOOF lodge on Briggs Street. John and his brother, J.J., opened a store together, but John eventually bought out J.J.'s half. He ran the store alone starting in 1882 and moved it to the IOOF building in 1897. John moved to Denver when he sold the house to Henry Carter in 1909. Carter was a Civil War veteran who came to Erie and became a grain buyer.

**Harris/Brennan/Oakley House, 405 Pierce Street.** Built in 1885, this house was the longtime home of John Harris, a coal miner from England. He bought the property from Sidney Dillon. John's daughter, Jane Harris Brennan Oakley, gained title to the property in 1919 and lived there until the 1970s.

**VanValkenburg Undertaking/Bracegirdle/Leinweber House, 525 Pierce Street.** This 1884 structure was built on land that transferred title from Sidney Dillon to several others before Richard van Valkenburg bought it in 1886. It served as Erie's mortuary, an overflow for school classes in 1908, the coroner's office and as a residence around 1910. It was referred to as the "coffin house," as coffins were built in the back of the building and sold in front. J.G. Trezise bought it in 1904; he was Boulder's undertaker and coroner from 1892 until around 1911. John Bracegirdle bought the house in 1919; he was a miner at the Puritan Mine. Maxine Leinweber bought the house in 1957, and it stayed in her family until at least 1991.

**Stevens/Taylor House, 555 Pierce Street.** This house is estimated to have been built somewhere between 1902 and 1908. Sidney Dillon owned the property originally, and the title changed at least three times before a house was built on it. John A. Freeman, a physician and drugstore owner, owned the land, and then Reverend Richard van Valkenburg owned it. Walter and Margaret Lister bought it later for investment and possible rental purposes.

Alfred Stevens, a mine owner, was the probable builder and first homeowner. He bought it in 1902. The title passed to Erie Bank in 1905. Bank president William Nicholson transferred title back to Stevens in 1906.

Frank Taylor bought the house in 1910, where he lived with his wife, Myrtle T. Taylor. Frank was a barber. Myrtle was an agent for the Union Pacific and the Chicago, Burlington and Quincy Railroads. Title records show that Frank and Myrtle shared title at 555 Pierce until 1944. Myrtle took the title of the house in 1968.

**Taylor House, 575 Pierce.** This house was built sometime before the 1940s. The property title transferred from the St. Louis, Denver Land and Mining Company to Thomas Griffiths in 1881. Griffiths and others sold it to Frank and Myrtle Taylor in 1944; it then transferred to Myrtle in 1946.

**Erie Town Hall, 235 Wells Street.** In 1883, a false-front building was erected as a town hall on Wells Street. In 1930, a new town hall was built at the same location. This early twentieth-century commercial building was the home of Erie's government. The building was constructed in stages.

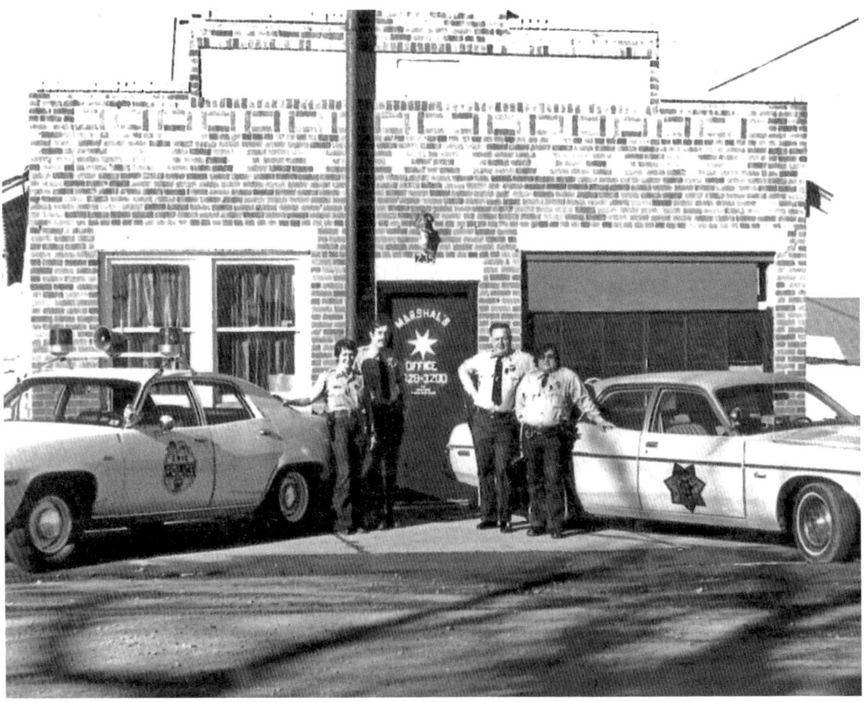


Erie's second town hall also housed its police department. *From left to right*: Donna Finley, Henry Aaronshaft, Chief Baker (with his .44 Hogleg) and Tony Montano, circa 1976. *James Sites collection.*

The first stage was built in 1930, with the front of the building housing offices, a meeting room and a bay for a fire truck for the Erie Volunteer Fire Department. The second stage was built at the rear in 1939 for the Erie Police Department and a jail. It also has been a voting place, draft registration site and the Erie senior center.

The building has unique features for Erie: a bell tower, a stepped parapet, a siren tower and contrasting brickwork. The structure was renovated in 2005 through a grant from the Colorado Historical Society. It currently is home to the Erie Chamber of Commerce.

The **Erie House** hotel was established at Wells and Kattell Streets. This was the first two-story building in Erie. For two dollars a day, miners could get a room, meals and use of the washhouse and stables. The building does not exist today, but the stables serve as apartments.

The **Valmont Cheese Factory** originated in 1877 in the Valmont area and was started by C.W. Hayden. The Valmont power plant was started by the Public Service Company in 1924 and served the Erie area.

A branch of the **Kuner-Empson Packing Company**, the Kuner-Empson pickle factory, operated at the south end of High Street. John Howard Empson was a Longmont industrialist who owned the Empson Packing Company in Longmont. Started in 1887, the company canned produce from Longmont area farms. Charles Lewis Hover bought the cannery in 1920 and, in 1927, merged it with the Kuner Pickle Company, owned by Max Kuner of Denver. It closed in the 1950s.

# 5
# FIRST CHURCHES

Many of Erie's early settlers had strong religious convictions and a major influence on the establishment of churches in the town. Before any churches were built, people worshiped in the schoolhouse, homes, stores and saloons.

Reverend Richard Jeptha van Valkenburg helped establish Erie's early Protestant churches. Van, as he was affectionately known, was ordained a Methodist minister in 1845 and served as a circuit preacher before settling in Erie around 1867 for health reasons. Besides preaching his way across the country, he conducted services in various Colorado towns, including Boulder, Longmont, Burlington, Valmont, Hagers and Upper St. Vrain.

Van was born on August 16, 1823, in Schoharie County, New York. He was the son of Edward van Valkenburg and Alice B. Grinnell van Valkenburg. Van married Cordelia Briggs on March 1, 1843, in Ulster, New York. Cordelia was born in December 1823, the daughter of Galen Briggs of Tioga, New York. They had four children—Martha Jane, Alice B., Galen B. and Mary E.—all born in New York and Pennsylvania. Cordelia died in Denver on January 22, 1914.

Van was a Civil War veteran. Holding the rank of captain, he raised a company of 108 men to serve in the Union army at the Battle of Antietam. Shortly after, he served as chaplain under General William Sherman on Sherman's famous march through Georgia.

Initially, Van preached and worked in the mines in Central City. He started Erie's first church and Sunday school in his living room. In 1883, he helped organize the Welsh Presbyterian church. In 1888, he had James Wilson build Erie's Methodist church.

He was a charter member and chaplain of the Erie lodges of the IOOF and the Masons. In the Odd Fellows, he served as Grand Chaplain, Grand Patriarch and Grand Representative to the State. He also founded the Good Templars. He served as mayor, town trustee, postmaster, police judge, president of the school board, justice of the peace and state legislator. He built the first house in Erie in 1872, established and operated the Erie House hotel and owned and edited the *Erie-Canfield Independent* newspaper. He acquired the house at 525 Pierce Street in 1886 and ran his own undertaking business until 1890.

Considered the "father of Erie," Van helped incorporate the town in 1874. He named it after one of his former residences, Erie, Pennsylvania. He died on September 8, 1912, and was buried in Fairmont Cemetery in Denver.

A Welsh Presbyterian church was built on Holbrook Street in 1883. Erie's large Welsh community wanted a place to worship. Referred to commonly as the Welsh Church, this was the first church built in Erie. Services were conducted in *Cymraeg*, the Welsh language. An advertisement from the *Erie-Canfield Independent* on May 30, 1890, lists John T. Williams as reverend. On September 16, 1898, the *Erie Review* listed Reverend R.J. Roberts as its pastor. Later, English residents wanted their own Presbyterian services and used the building.

**Erie United Methodist/Episcopal Church, 602–604 Holbrook Street.** The Methodist church was organized in 1872, but the members met without a dedicated site. James Wilson, a builder and brick maker, was asked to build the late Victorian/Gothic Revival–style church for Erie's Methodist community in 1888. When the congregation outgrew the building, the Pleasant View church was moved into town and annexed to the original church. Around 1920, when the congregation again outgrew this facility, the Presbyterian church was moved up Holbrook Street and annexed to the existing church. Later, this annex became the Erie Community Center. Some claim the social hall was burned to the ground in 1951, and the area was filled in for a parking lot. Marjorie Smith recalls serving turkey dinners in the basement of the hall in the 1980s. Today, however, the site *is* a parking lot for the Methodist church.

The church has several stained-glass windows. One was made and installed in memory of Erie's Webber brothers—John, Thomas and William—who died in Utah's Schofield Mine in 1900. Another is dedicated to Reverend van

Valkenburg. Still others bear the names of David F. Davis, Sadie J. Mitchell, W.J. Keeser, William Nicholson, J.J. Wharton, Edgar F. White, Reverend Joel F. Smith and Reverend Thomas Hooper.

Other features memorialize early residents: the communion table for Thomas P. Morgan (1851–1930) and Sarah Ann Morgan (1851–1922); the lectern for Joseph Butcher (1845–1922) and Maria Butcher (1849–1925); the transom for William Nicholson (1850–1933); the offering plates for Adair C. Winslow (1903–1925) and Susana Elizabeth Haley (1856–1920; the cross for the Epworth League, a young adult association; and the chancel hangings for the Ladies' Aid.

The congregation was originally led by Reverend van Valkenburg, although he was never recognized by the church as being the assigned minister. Officially, a minister from a nearby town was the registered reverend. Following is a list of the church's ministers:

| MINISTER[1] | YEARS | MINISTER | YEARS |
|---|---|---|---|
| Richard J. van Valkenburg | 1867–1870[2] | Oscar F. McKay | 1886–1889 |
| George Wallace | 1871–1872 | Thomas Andrew | 1889–1891 |
| John L. Dyer | 1873 | Hosea L. Beardsley | 1891–1892 |
| Isaac H. Beardsley | 1874 | Kent White | 1893 |
| John L. Moffat | 1874–1875 | Gustavus A.W. Cage Jr. | 1894–1895 |
| George Wallace | 1875 | Edgar White | 1895 |
| John Collins | 1876 | Henry J. Grace | 1896 |
| A.N. Field | 1877–1880 | A. Coleman | 1897 |
| A. Coleman | 1881 | Jacob R. Rader | 1897–1898 |
| A.D. Fairbanks | 1882 | Thomas J. Hooper | 1899–1901 |
| R.L. Kenyon | 1883 | John C. Gullette | 1902–1903 |
| Reverend Harrington | 1883–1884 | Robert McLaren | 1903–1904 |
| John A. Long | 1884–1885 | R.A. Carmine | 1905 |
| T.A. Storey | 1885 | Arthur Morrison | 1905–1907 |

| Minister | Years | Minister | Years |
|---|---|---|---|
| W.J. Richards | 1907–1910 | J.E. Carothers | 1933–1937 |
| John Oliver | 1910–1911 | Fred Clark | 1937–1938 |
| Dougald C. Winship | 1912–1913 | Leo J. Wolfe | 1939–1940 |
| C.L. Marsh | 1914–1916 | F.A. Napier | 1941–1942 |
| C.M. Kennaugh | 1917 | G.W. Conklin | 1943–1944 |
| F.A. Mead | 1917 | Don Littrell | 1944 |
| Dougald C. Winship | 1917 | Reverend Pulliam | 1944 |
| W.G. Scott | 1917 | Richard Evans Brown | 1945–1946 |
| Frank F. Gibbs | 1917–1918 | Reverend Fisher | 1946 |
| Chester Jennings | 1917 | Earl Lewis | 1947–1950 |
| J.A. Moorman | 1917 | Earl Keith Hanna | 1951 |
| W.A. Keeser | 1917 | Bruce E. McConnell | 1952 |
| C.B. Boddy | 1917–1918 | Earl Keith Hanna | 1953 |
| W.J. Keeser | 1918–1922 | Raymond Ault Thompson | 1953–1954 |
| R.S. Jones | 1922 | Manford L. Jackson | 1955 |
| T.M. Davis | 1922 | Allen Keith Jackson | 1956–1958 |
| H.G. Honnold | 1923–1924 | no name listed | 1959–1960 |
| M.T. Habgood | 1924 | Kenneth G. Moreland | 1961 |
| W.J. Keeser | 1925–1927 | Robert L. Hamilton | 1962–1963 |
| W.E. Tinney | 1928–1929 | Richard Halbert | 1964 |
| A.B. Horton | 1930 | Kenneth Thurman | 1964 |
| F.H. Rose | 1931–1933 | Tom English | 1964–1965 |

| Minister | Years | Minister | Years |
|---|---|---|---|
| Robert T. Lynch | 1965–1966 | Richard R. Lippincott | 1984–1994 |
| Robert Lavern Alcott | 1966–1968 | Dave B. Lillie | 1994–1996 |
| Gale D. Boockes | 1969–1970 | Dave Moorman | 1996–1998 |
| Joe E. Wood | 1970–1972 | Dave L. Olson | 1998–2000 |
| Grover Briggs | 1973 | Ronald Kowalski | 2000 |
| Harold T. Urton | 1973–1975 | Dale Lovett | 2000–2001 |
| no name listed | 1976 | Aaron Gray | 2002–2004 |
| R.B. Bryant | 1977–1983 | Hugh Harris | 2005–2006 |
| Alexis Sidorak | 1984 | Jim Calhoun | 2006–2014 |
| Kenneth Wright | 1984 | Carol Lillie | 2014–present |

1 This list is incomplete, as many others served the Methodist community over the years; however, records have been lost or people never were recorded as having served. Much of the information on the church comes from an unpublished manuscript from 1998 by Reverend Dave L. Olson entitled "A History of the Erie United Methodist Church." Additional information was made available by Reverend Jim Calhoun.

2 R.J. van Valkenburg was not listed officially as the church minister, but he served Erie's Protestant/Methodist population during this time.

For a more detailed history of the church, visit Erie Methodist/Episcopal Church or visit its website at www.erieumc.org.

**St. Scholastica Roman Catholic Church, 575 Wells Street.** Erie's earliest Catholic mass is recorded as having been in 1877, offered by traveling priests from Pennsylvania. By 1890, regular masses were provided every third Sunday by priests from towns near Erie.

Mike Brennan hauled a recycled barn in from Canfield to serve as Erie's first Catholic church building. It was expanded, remodeled, updated and enhanced as the congregation grew.

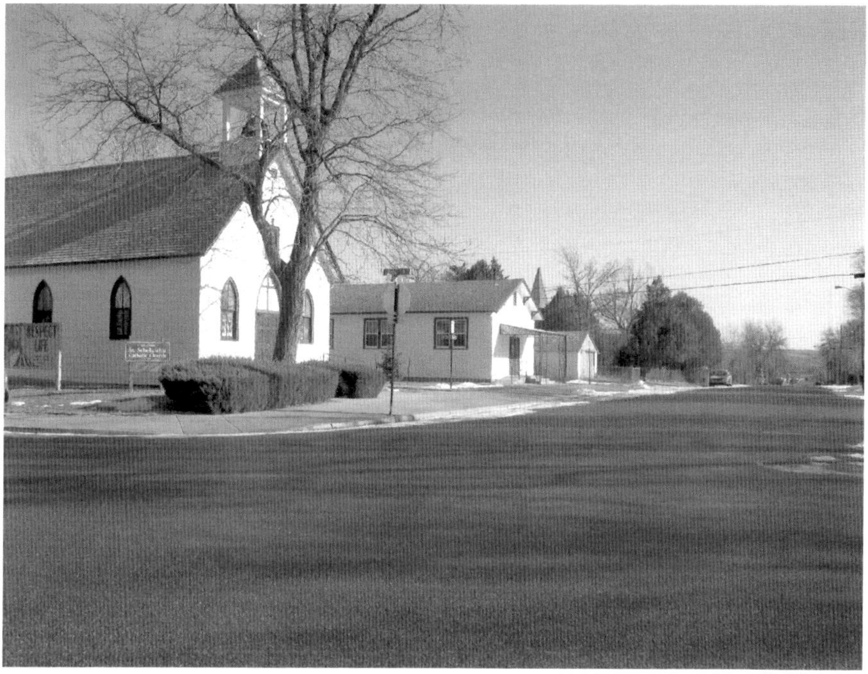

The St. Scholastica Catholic Church and social hall. *Author's collection.*

Erie's Catholic community helped to found and build St. Scholastica as a missionary church in 1898 at 575 Wells Street. Lithuanians, French, Slavic and Italian miners all pitched in to build the church. Father Cornelius Enders of the Order of St. Benedict (OSB) oversaw this effort. He had served earlier in Ward, Boulder and Longmont. He was born on December 7, 1866, to Johann and Johanna (Schaller) Enders in Kennersreuth, Bayern, Bavaria, Germany. He entered the novitiate in 1886 and took his vows in 1890. Before coming to Erie, he had presided over masses and the building of churches and schools in various towns.

Benedict of Nursia (Italy) was born around 480 AD. As a young man, he was sent to Rome to finish his studies. At that time, the Roman Empire was being overtaken by Goths and Vandals, who were corrupt and secular. Benedict did not feel comfortable in Rome and feared for his life. He decided to devote his life to spiritual work and entered a countryside monastery called Montecassino.

Because Benedict showed initiative about reform, he was ostracized by other monks. In time, he recruited others who believed in his philosophies, which became known as *The Rule of St. Benedict*. In short, he advocated the renunciation of worldly pursuits and the devotion of one's life to spiritual work; this is the foundation of Western monasticism. *The Rule of St. Benedict* was one of the most important written works used to shape medieval Europe. Charlemagne regarded it as a model for other religious orders to follow, and he championed this movement. Benedict founded the Benedictine Missionary Movement and eventually the Order of St. Benedict, which attracted many followers because his outlooks provided a balance between individual zeal and strict fundamentalism.

St. Benedict is the patron saint of numerous causes, against a variety of diseases and inflictions, for a number of trades and professions and particularly the patron saint of monks and of Europe. His name means "blessed." He died around 547 AD.

Scholastica was Benedict's twin sister. She was a nun and the leader of a women's community called Plumbariola, about five miles from Montecassino.

Once a year, Scholastica and Benedict would meet outside Montecassino to discuss a variety of religious topics and various sacred texts. Some report that she may have received her calling to God before Benedict received his. She founded a women's branch of Benedictine Monasticism.

St. Scholastica is the patron saint of nuns, convulsive children and against rain and storms. She symbolizes "she who has leisure to devote to study." Her memorial day is February 10, the date she is reported to have died in 542 AD.

Reverend Enders and his parishioners brought much of the lumber from the mountains by wagon to build St. Scholastica. On Sunday, August 18, 1899, St. Scholastica was dedicated. Worshipers came from Lafayette, Longmont, Louisville and Boulder to participate in the event. In 1909, Mr. and Mrs. George Zimmerman donated a bell to be installed in the church steeple. Various donors—including Jerry Kincade, Marion Greely and Mary Miller—are responsible for all of the stained glass windows. Some were installed around 1976 and others around 1997. A parish convent was acquired in 1999.

In 1927, it was decided that a church was needed in the mining camp of Serene at the Columbine Mine. Someone purchased a deserted frame church in Louisville and moved it to the mine. It served both as a church and the post office from 1927 to 1931. In 1931, services were no longer offered at the mine. In 1940, the building was moved to Lafayette to serve as the parish hall at St. Ida. In 1957, St. Ida needed a larger parish hall, and the building was moved to Erie for St. Scholastica. While remodeled and updated considerably, the frame still serves as the parish hall today.

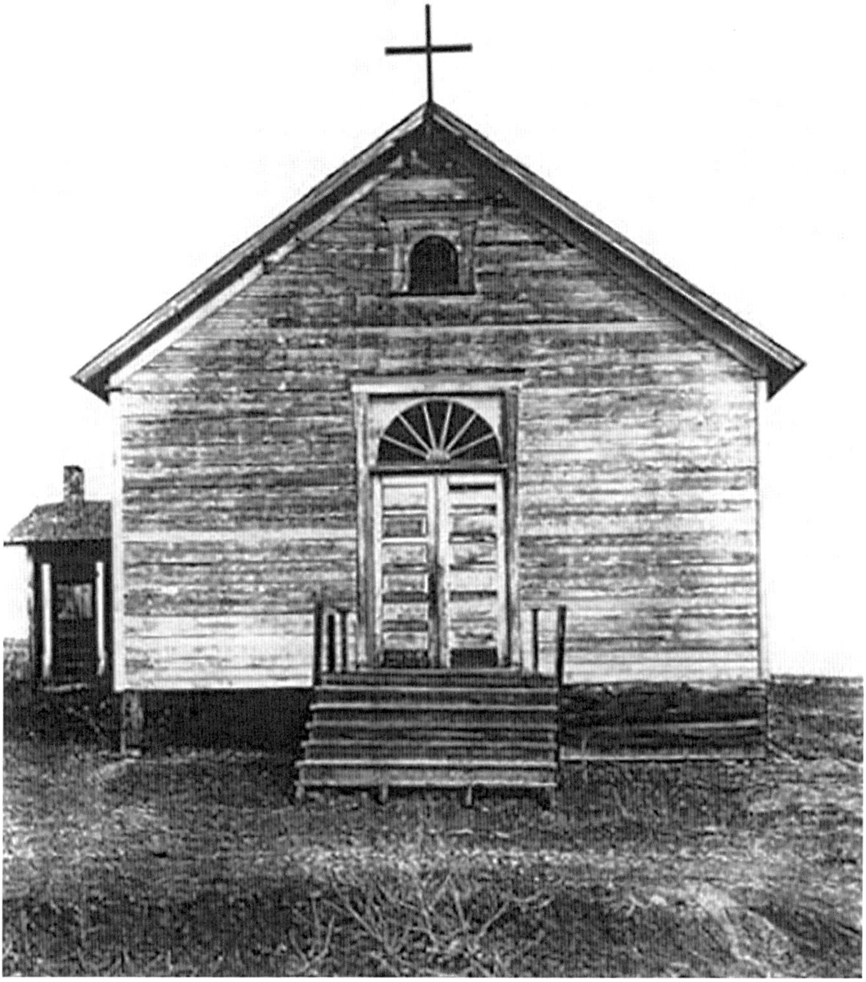

This church was brought from Louisville, to Serene, to Lafayette and finally to Erie. *St. Scholastica collection.*

In 1931, St. Scholastica became a mission church of St. Ida's in Lafayette. In 1975, it became a mission church of St. Theresa's in Frederick.

In 1988, plans were drawn to enlarge and beautify St. Scholastica. Parish volunteers remodeled the ceilings, enlarged the interior and installed new pews to make the structure something parishioners enjoyed with pride. On July 28, 1991, the church was rededicated by Archbishop J. Francis Stafford.

The following table lists the leaders of the Catholic church in Erie:

| Name | Dates | Name | Dates |
|---|---|---|---|
| Reverend R.M. Goodman, OSB | 1890–1892[1] | Reverend Raymond Layton, OSB | 1942 |
| Reverend Homan, OSB | 1893[1] | Reverend Maurus Zabolitsky, OSB | 1943–1946 |
| Reverend Cornelius Enders, OSB | 1898[1]–1899[2] | Reverend Paulinus Hammer, OSB | 1946–1947 |
| Reverend Lawrence Haas, OSB | 1899–1900 | Reverend Maurus Zabolitsky, OSB | 1947–1949 |
| Reverend Nicholas Siedl, OSB | 1900–1902[3] | Reverend Claude Roberts, OSB | 1949 |
| no names listed | 1902–1908[4] | Reverend Joseph Hannan, OSB | 1949–1956 |
| Reverend Suitbert Rickert, OSB | 1908–1910 | Reverend Vincent Peter, OSB | 1956–1957 |
| Reverend Antonine Hintenach, OSB | 1911–1914 | Reverend Martin Arno, OSB | 1957–1960 |
| Reverend Robert Murray, OSB | 1914–1919 | Reverend Joseph Hannan, OSB | 1960–1961 |
| Reverend Regis Barrett, OSB | 1920–1926[5] | Reverend Dennis Langdon | 1961–1964 |
| Reverend Albert Schaller, OSB | 1926–1931[6] | Reverend Peter Hudson | 1964–1966 |
| Reverend Robert Murray, OSB | 1931–1937[7] | Reverend Martin Arno, OSB | 1966–1968 |
| Reverend Urban Schnitzhofer, OSB | 1937–1941 | Reverend Michael Walsh | 1968–1970 |
| Reverend Matthew Connelly, OSB | 1941 | Reverend Kenneth Q. Funk | 1974 |
| Reverend Charles Forsyth, OSB | 1941 | Reverend David Sobieszczk | 1974 |

| Name | Dates | Name | Dates |
|---|---|---|---|
| Reverend James Morgan | 1974–1975[8] | Reverend Luis Garcia, J. C. L.[9] | 1991–1995 |
| Reverend Peter Urban<br>Reverend Leonard Urban | 1975–1986 | Deacon Daniel Marthe | 1992–1995 |
| Reverend John D. McCormick | 1986–1988 | Very Reverend G. Timothy Gaines, V.F.[10] | 1995–2000 |
| Reverend Donald C. Willette | 1988–1991 | Reverend Robert Whipkey | 2000–2007 |
| Deacon Martin L. Sanchez | 1989–1992 | Reverend Hernan Florez | 2008–present |

1 Masses were held at the Erie Public School.

2 Masses were held from this date on at St. Scholastica.

3 During this time, St. Scholastica was a mission of Longmont's St. John the Baptist.

4 Masses were conducted by reverends from Sacred Heart of Mary in Boulder from 1902 to 1926.

5 Approximate dates

6 Based on baptismal records

7 From 1931 to 1975, St. Scholastica was a mission of St. Ida's Church in Lafayette. Services were provided by Benedictine *and* diocesan priests.

8 From 1975 until 2007, St. Scholastica was a mission of St. Theresa in Frederick. Apparently, there is still some discussion about whether it is still a mission church.

9 *Juris Canonici Licentiatus* or Licentiate in Canon Law

10 V.F. is *Vicar Forane*. Very Reverend signifies a specialized role.

Much of this information was provided by unpublished manuscripts written by parishioners of St. Scholastica: "St. Scholastica Church 1899–1999" (Jane Kozma, Jane Marshall, Mary Montano and Dave Stone) and "St. Scholastica Church 1899–2009" (Jane Kozma, Jane Marshall, Mary Montano, Mary Dieke and Dave Stone).

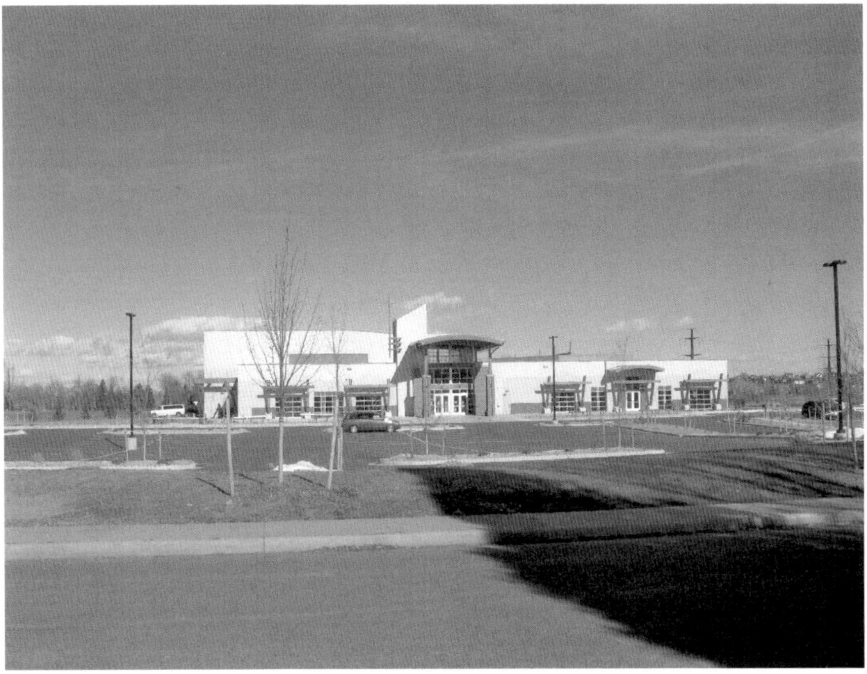

The Calvary Bible Church moved into its own building in 2014. The congregation used to meet at Erie High School. *Author's collection.*

For a more detailed history of the church, contact St. Scholastica or visit its website.

6

# FIRST SCHOOLS

Most of the early records of Weld County school districts have been lost. The following information has been pieced together from books and interviews.

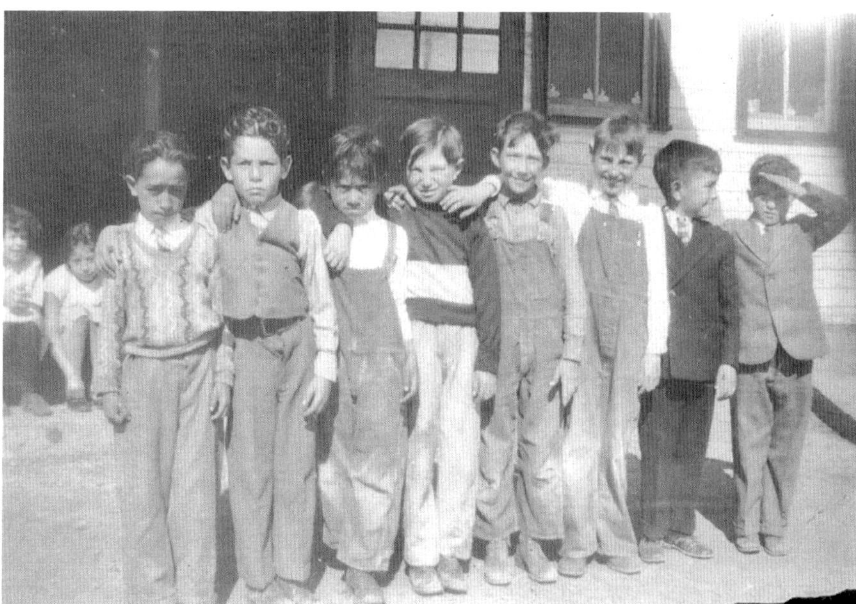

A group of boys line up for a picture at the school at the Columbine Mine around 1927. *Lafayette Public Library.*

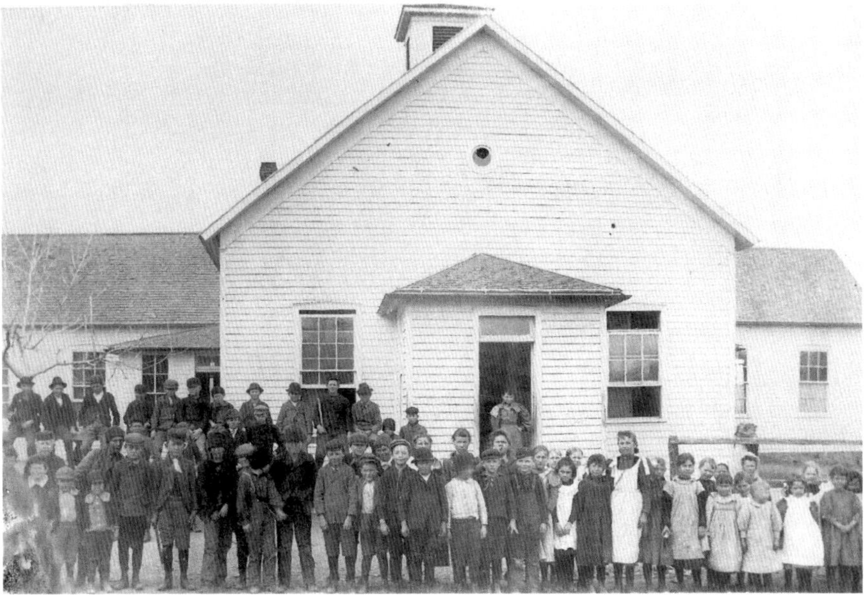

Erie's first school was located on the southeast corner of Briggs and Wells Streets. *Erie Historical Society.*

The first school near Erie was the Idaho Creek School, located south of Highway 52 and east of East County Line Road. It was founded by Robert Block, David Irwin and Morse Coffin on September 1, 1861. Today, Idaho Creek is listed as a populated place in Weld County on Erie's Geological Survey and located northeast of Erie, north of Highway 52.

Erie's original schoolhouse was built in 1881 on Briggs and Wells Streets, where today's United States Post Office exists. It is believed that it served over one hundred students. The first teacher in Erie was Mary L. Young.

**Lincoln School, 645 Holbrook Street.** The next schoolhouse, the Lincoln School, was built on Holbrook and Wells Streets, with one large room in the center and one each on the south, west and east. It was made of bricks. Classes were held there in 1907. In 1908, overflow students were taught at the "coffin house" at 525 Pierce Street, where a mortuary and undertaker's business was located. The brick building was eventually deemed inadequate, so in 1920, four additional rooms were built on the north end of the building. Lincoln School served grades one through twelve. In 1929, students in grades seven through twelve moved to the new high school. In 1966, Erie Elementary School was built on County Line Road, and the Lincoln School building was abandoned. It was later given to the Town of

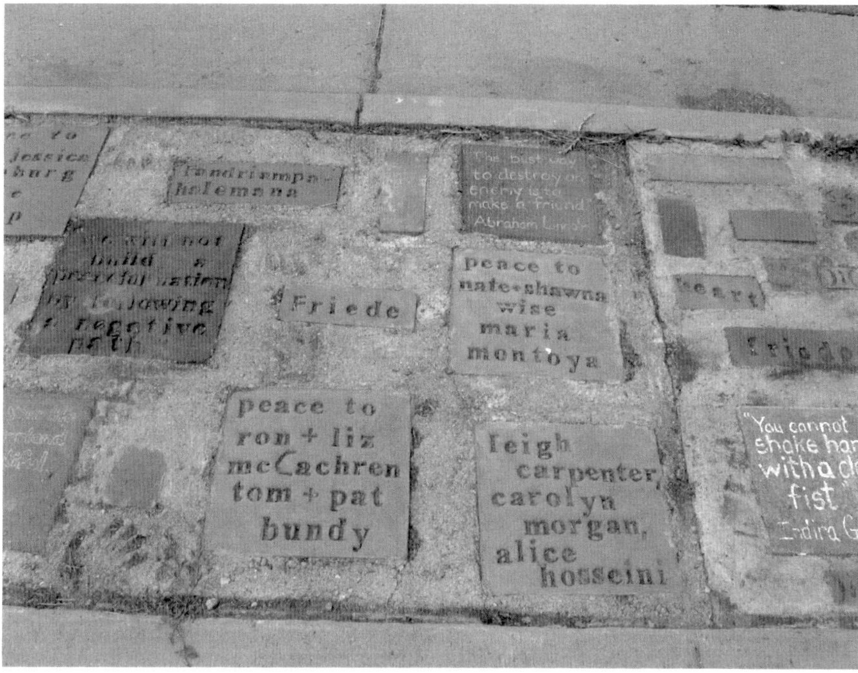


Students from Erie Elementary School were asked to make pavers to reflect their feelings about the 9/11 attacks on the World Trade Towers. Pavers are also displayed at the Arts Coalition of Erie and the Erie Community Library. *Author's collection.*

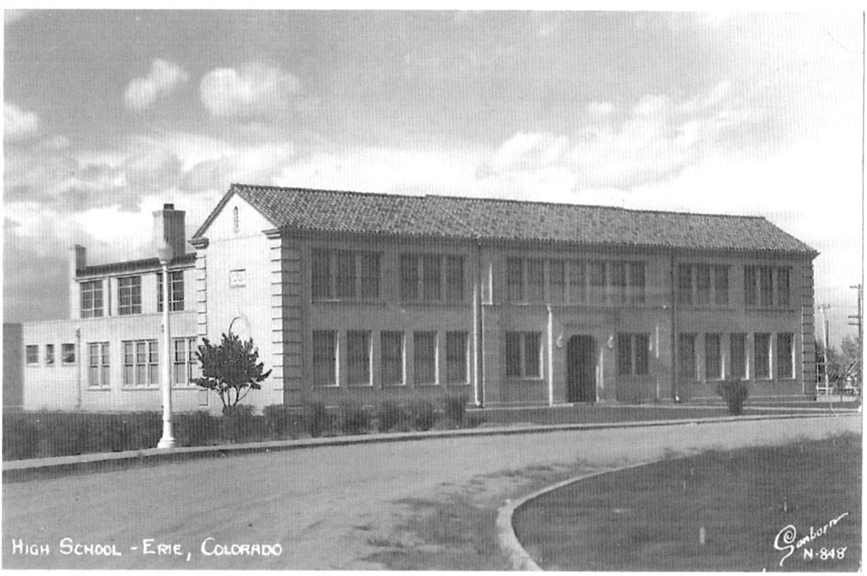


Erie High School opened in 1929 to accommodate grades seven through twelve. *Ray Wilson collection.*

Erie by the school district and has been used by the town government since. The building is on the National Register of Historic Places.

**Erie High School, 650 Main Street.** In 1929, a new school was constructed on Main Street at Cheesman to serve grades seven through twelve. Its western boundary is East County Line Road.

While there is a paucity of information on the school, documents provided by Debbie Langerak show that in 1945, Carmella Domenico and H.C. Shiffy were on the faculty. The Parent-Teacher Association (PTA) program for the 1954–55 school year lists the Erie High School principal as Charles Peccolo and the grade school principal as Rae Prickett. Faculty listed for that year were Elaine Andre and Eleanor Simmons (grade one), Bertha Jones and Daisy Adair (grade two), Arlene Mercer (grade three), Ava Giles (grade four), Stella Lee (grade five), Marvel Walker (grade six), Marion Berges (grade seven), Charles Simon (grade eight), Elizabeth Ozment (grade nine), Glen Bruning (grade ten), Harold Conroe (grade eleven), Joan Smith (grade twelve), Barbara Carlson, Dave Lewis and Alice Peterson.

The PTA program for the 1955–56 year lists the Erie High School principal as Gustave Blade and the grade school principal as Rae Prickett. Faculty are listed as Irma Bledsoe and Ellen Regnier (grade one), Daisy Adair and Bertha

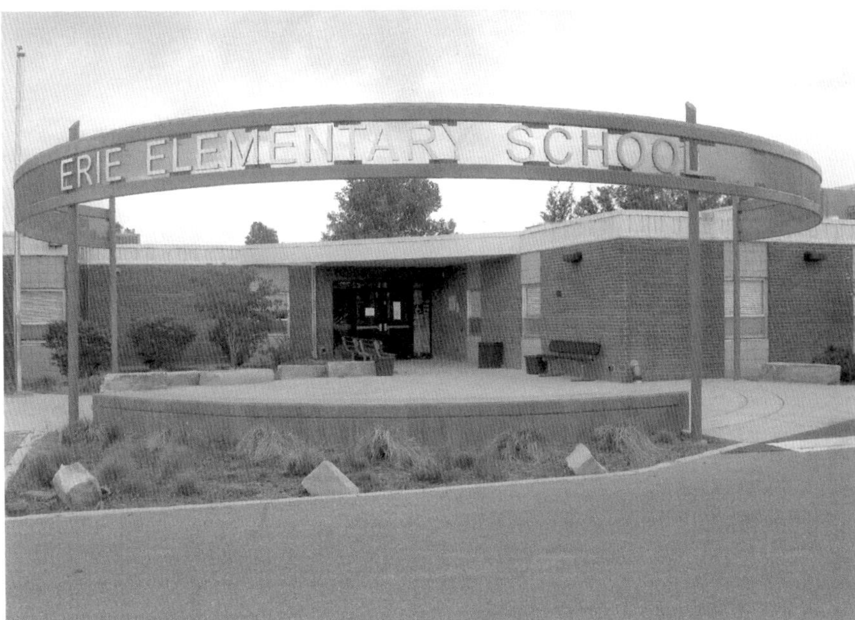


Erie Elementary School was built in 1966 on East County Line Road. *Author's collection.*

Jones (grade two), Arlene Mercer (grade three), Elna Bacon (grade four), Randall Ruff (grade five), Marvel Walker (grade six), Arthur Wright (grade seven), Doris Prindle (grade eight), Jack Salo (grade nine), Jeanne Sible (grade ten), John Tingstad (grade eleven), Harold Conroe (grade twelve), Richard Bachman, Alma Brennan, Sarah Dodge and Dean Simpson.

Some of the town businesses that placed advertisements in the *Erion*, the school's newspaper, were Taylor's Barbershop, Ann's Place, Bob's Café, Walt's Tavern, Tommie's Service Station, Wilson Larson, Hogsett Lumber Mercantile Company, Winslow's Store and J.H. McDonald Insurance Agency.

**Erie Elementary School, 4137 East County Line Road.** Erie Elementary School was built across from Erie High School in 1966 for grades one through six. The first principal of the new school was Ivan Adams, who had been the principal at the Lincoln School. The brickwork was done by the M.W. Grand Lodge A.F. and A.M. of Colorado.

# 7
# MINES

Coal falls into four categories, from highest to lowest quality: anthracite, bituminous, subbituminous and lignite. Lignite coal—found in the Boulder-Weld Coal Field—is a soft, brownish-black coal that contains moisture and has a carbon base. It is unstable and can combust during lengthy storage or transportation. It turns to powder if exposed to air for a long time. It was used primarily for train fuel or steam-electric power. Underground, it can burn for a long time; mine fires have been discovered as recently as the 1980s, and still today, people report seeing smoke coming out of the coal seams near Superior and Marshall.

The Northern Colorado Coal Field extends twenty to twenty-five miles from Marshall in the southwest part of the field through Louisville, Lafayette, Niwot, Erie, Eastlake (Thornton, near 128th and Claude), Dacono, Frederick, Firestone and Gowanda (east of Interstate 25 on Highway 66). The area encompasses 6,800 square miles. While the greater coal field contains nearly 200 abandoned mines, the Wise Homestead Museum lists about 140. One of the reasons for so many individual mines is because new mines were often dug to avoid having to drill through the many rock faults in the field. This chapter covers the 47 associated with Erie.

The Arapaho and Cheyenne found coal on the ground and marveled at its ability to burn for a long time, but they didn't capitalize on its abundance in the area or adopt it into their cultures. Many early settlers came from locations where coal was common; they picked up surface coal and easily tapped shallow veins for coal to be burned to heat their homes. The earliest official discovery of coal in the Northern Colorado Coal Field was in Marshall in 1856.

Jim Baker dug the first coal mine in Erie. It was never listed officially, probably because it was taken over by the larger Boulder Valley Mining Company. He began coal mining around 1859, clearly before the official discovery of coal in Erie in 1866. George Gilson, a homesteader and slope miner, realized his land contained a rich coal source on the hillside east of Erie. He traded the land to Captain Ira Austin.

In 1868, Henry Briggs, a fifty-two-year-old coal miner from New York, received a patent to work 160 acres on the east edge of Erie. The Briggs Mine opened in 1871 and was the first commercial coal mine in the Erie area. It had four coal veins and about five hundred feet of sloping shaft. It was also the largest producer of coal in the Northern Colorado mining district for several years, responsible for about 70 percent of the area's coal output. The Boulder Valley Coal Company eventually acquired the Briggs Mine and renamed it the Boulder Valley Mine. The Boulder Valley Coal Company operated seven mines, including the Rob Roy and the Star. John McCool came to Erie from Denver, where he had been a boot and shoe merchant. In 1871, McCool was granted a coal mining patent, signed by President Ulysses S. Grant. He worked the western 160-acre segment of the same quadrant worked by Briggs. The middle 320 acres of the quadrant went to Ira and Schuyler D. Austin.

It had become clear that the area around Erie was rich with coal. It would become necessary to dig deep shafts underground to locate many of the coal seams and mine them. According to the United States Geological Survey (USGS), coal beds in this area were found near the surface and deeper than four hundred feet. Tunnels were dug as far as a mile and a half in all directions from the original mine heads. The tunnels were supported by wooden timbers. Charlie Waneka of Lafayette reports that many of the mines were connected. In poor weather, people would sometimes walk for miles in this protected underground system to get from one place to another.

Erie and its surrounding area became the largest-scale producer of coal in the Northern Colorado Coal Field. Soon, large mining companies and railroads moved into the area to claim land and open mines. Mining became the most significant industry in Colorado in the nineteenth and early twentieth centuries. Some of Erie's mines were located west and southwest of town; others were east of Coal Creek. Erie was the most important coal town in Colorado. Following is a list of the mines.

Many mines were dug next to Old Town Erie, and people often ask where they were. In some cases, remains offer clues; in others, only a USGS map can help pinpoint locations. In the following sections, the mines are listed by vicinity, name, dates of operation (if known) and approximate location. The last operating mine in the Erie area was the (new) Lincoln Mine, east of Interstate 25. Before it closed in 1979, it produced about five hundred tons of coal per day.

| Mine | Dates |
|---|---|
| Briggs Shaft/Mine (bought by Boulder Valley) | 1871–???? |
| Mitchell | 1883–1891 |
| Stewart | 1883–1892 |
| Northrup | 1883–1894 |
| Star | 1883–1895 |
| Boulder Valley (old) | 1883–1885 |
| McGregor | 1884–1895 |
| Garfield #1 | 1884–1897 |
| Cleveland | 1885–1895 |
| Jackson | 1885–1890 |
| Garfield #2 | 1886–1905 |
| Boulder Valley (old—reopened) | 1890–1893 |
| Northwestern | 1892–1907 |
| Lincoln (old) | 1896–1907 |
| Chase | 1897–1895 |
| Long's Peak | 1897–1901 |
| Marfel | 1897–1904 |
| Tynon | 1900–1904 |
| Lehigh | 1902–1910 |
| Reliance | 1903–1907 |
| Washington (old) | 1903–1911 |
| Crescent | 1905–???? |

## Four Corners Area

*Intersection of Erie Parkway and East County Line Road*

**Briggs Shaft/Briggs Mine.** The closing date of the mine's operation is unknown. Briggs received a patent to work the area in 1868 but actually opened the mine in 1871. It was east of Coal Creek. Miners settled near the mine at the Parkdale junction just northeast of Erie. In 1871, the Union Pacific Railroad tracks reached Erie. Previously, coal had to be hauled seven miles from the Briggs Mine to the existing railroad tracks in Hughes (Brighton). The Boulder Valley Coal Company eventually bought and operated the mine.

**Garfield No. 1.** 1884–97. Garfield No. 1 was located on the

| Mine | Dates |
|---|---|
| Parkdale | 1907–1916 |
| State | 1913–1916 |
| Washington (new) | 1915–1918 |
| Boulder Valley #1 | 1917–1947 |
| Star (new) | 1918–1921 |
| Clayton | 1920–1942 |
| Columbine Hoisting Shaft | 1920–1946 |
| Columbine Rock Slope Entry | 1920–1946 |
| Imperial | 1927–1972 |
| Morrison | 1930–1953 |
| Banner Shaft | 1932–1941 |
| Arrow | 1933–1935 |
| Munroe | 1933–1940 |
| Eagle | 1939–1978 |
| Baseline | 1941–1942 |
| Lincoln (new) | 1948–1979 |
| Lloyd Shaft | unknown |
| Pinnacle | unknown |
| Canfield | unknown |
| Standard | unknown |
| Standard | unknown |
| Standard | unknown |
| McKissick | unknown |
| Unnamed (near Grandview) | unknown |
| Unnamed (near Northridge) | unknown |
| Unnamed (near Northridge) | unknown |

northwest corner of the Four Corners intersection on Erie Parkway. You can still see the tailing pile there, which is the residue left over when separating the coal from the dirt and rock.

**Garfield No. 2.** 1886–1905. The Garfield No. 2 mine was located north of Erie Parkway and east of East County Line Road, under the Erie Community Park. City officials report that this mine was so near the surface that no planting will be done along County Line road outside the tennis courts for fear that the ground will subside into old mine shafts.

**Lloyd Shaft.** The exact dates of operation for this mine are unknown. The mine was located south of Erie Parkway across the street from the Erie Community Center parking lot, between the small shopping center and Walgreens Drugs. Lloyd Avenue runs directly into the shopping center.

**Marfel Shaft.** 1897–1904. This mine was located on the

southwest corner of Erie Parkway and East County Line Road in an open field across from Walgreens Drugs and in front of St. Luke Orthodox Christian Church.

**Cleveland.** 1885–95. The Cleveland Mine was on the west side of East County Line Road, just south of the St. Luke Orthodox Christian Church. A dirt road still exists on the west side of County Line Road that goes to the mine area. West of the mine is a neighborhood accessible from Austin Avenue.

**McGregor.** 1884–95. The McGregor Mine was located on the west side of East County Line Road, just southwest of the parking lot behind the St. Luke Orthodox Church.

**Pinnacle Shaft.** The exact dates for this shaft are unknown. The shaft was located in an open field just west of the Four Corners intersection on Erie Parkway on the northwest corner, just east of Garfield No. 1.

**Mitchell.** 1883–1991. The Mitchell Mine was located east of East County Line Road at the intersection of Mitchell and Lloyd Streets on the northeast corner.

## Erie Commons

**Lehigh.** 1902–10. The Lehigh Mine was located in Erie Commons east of the 1000 block of Zodo Avenue, south of Austin just past the cul-de-sac. Bryan and Susan Holland saw excavated mine timbers and shafts being unearthed as the land was prepared for new homes on Nelson Avenue.

## Telleen Avenue and East County Line Road

**Stewart.** 1883–92. The Stewart Mine was located on Jackson Drive west of East County Line Road and south of Telleen Avenue, near where Jackson Drive turns and runs east–west.

**Northrup.** 1883–84. The Northrup Mine was located west of East County Line Road, south of the railroad tracks and north of Telleen Avenue.

**Chase.** 1892–95. The Chase Mine was located west of the intersection of Telleen Avenue and Meller Street, near where Red Hawk Elementary School was built. It is accessible from a dirt road off Meller.

## Jay Road and East County Line Road

**Long's Peak.** 1892–1901. The Long's Peak Mine was located east of East County Line Road and south of Jay Road between North 123rd and North 124th Streets.

**Star.** 1883–95. The Star Mine was located north of Jay Road and west of North 119th Street. A dirt road and Boulder Creek mark the entrance to the mine area.

**Star (new).** 1918–21. The new Star Mine was located north of Jay Road just after it makes the S curve going west. It was near the Arrow and Crescent Mines.

**Arrow.** 1933–35. The Arrow Mine was located north of Jay Road just after it makes the S curve going west. It was near the new Star and Crescent Mines.

**Crescent.** 1905. The Crescent Mine was located north of Jay Road just after it makes the S curve going west. It was near the new Star and Arrow Mines.

**Canfield Shaft.** The exact dates for this shaft are unknown. The Canfield Shaft was located west of East County Line Road and Long's Peak Mine, south of Jay Road and south of North 123rd Street.

**Standard.** The exact dates for this mine are unknown. The Standard Mine was located on an unnamed road west of North 123rd Street and north of the railroad tracks. This was near the location of the other listed Standard Mines.

**Jackson.** 1885–90. The Jackson Mine was located west of North 119th Street, south of Jay Road and north of Erie Parkway. If Billings extended past North 119th Street to the east, it would run just south of the mine, which is in the middle of a large field.

**Banner Shaft.** 1932–34. The Benner Shaft was located north of Billings Avenue and west of North 119th Street.

## Jasper Road and Canfield Area

**Standard.** The exact dates for this mine are unknown. This Standard Mine was located south of Jasper Road and west of North 119th Street.

**Standard.** The exact dates for this mine are unknown. This Standard Mine was located south of Jasper Road and west of North 119th Street, farther west than the other Standard Mines.

**Tynon.** 1900–1904. The Tynon Mine was located south of Jasper Road, east of North 119th Street, south of Flatiron Drive and northeast of Tynan Drive (note the spelling difference).

## Erie Air Park and Bonanza Drive

**Baseline.** 1941–42. The Baseline Mine was located on Bonanza Drive/County Road 3, just south of Sunset near Erie Airpark and Highway 7. This is probably an example of different shafts being dug into the same vein.

**Munroe.** 1933–40. The Munroe Mine was located on Bonanza Drive/County Road 3, just south of Sunset near Erie Airpark and Highway 7.

**Washington (new).** 1915–18. The new Washington Mine was located on Bonanza Drive/County Road 3, just south of Sunset near Erie Airpark and Highway 7.

**Washington (old).** 1903–11. The old Washington Mine was located on Bonanza Drive/County Road 3, just south of Sunset near Erie Airpark and Highway 7.

**Parkdale.** 1907–16. The Parkdale Mine was located on East Baseline Road, just east of Airport Drive entering Erie Tri-County Municipal Airport.

**Lincoln (old).** 1896–1902. The old Lincoln Mine was located on Bonanza Drive/County Road 3, just south of Sunset and north of Highway 7.

## Erie Parkway and Weld County Road 5

**Columbine Hoisting Shaft.** 1920–46. The Columbine Hoisting Shaft was located in the fields south of Erie Parkway and west of Weld County Road 5. If you look south from the stop sign by Erie High School, you will see a covered tailing or slag pile from the state mine. If you drive south on Weld County Road 5, you will notice a road leading into the Front Range Landfill. The road takes you close to the mine. The three other mines also located in that area and are listed on page 78.

*Opposite, bottom*: Miners pose in front of the tipple at the State Mine in 1927. *Lafayette Miners Museum.*

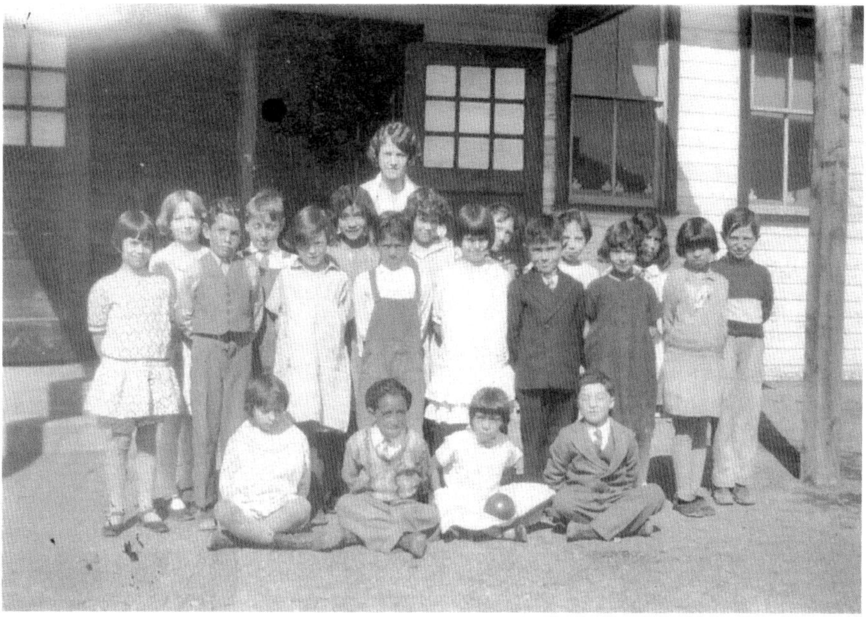

This unidentified teacher and her class pose for a picture at the Columbine Mine school at Serene around 1927. *Lafayette Public Library.*

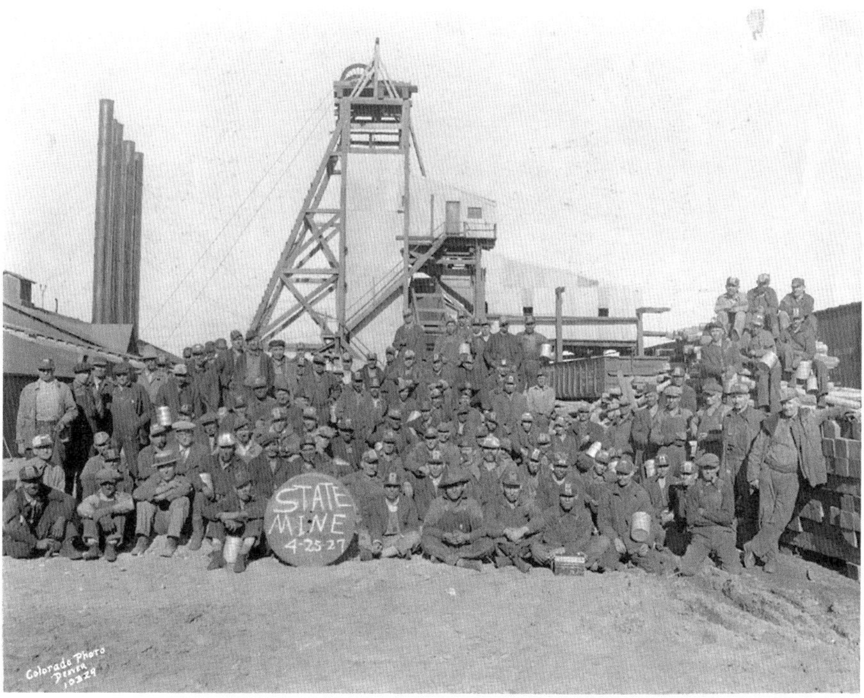

**Columbine Rock Slope Entry.** 1920–46. The Columbine Rock Slope Entry was located out in the fields just south of Erie Parkway and west of Weld County Road 5.

**Boulder Valley Number 1.** 1917–47. Boulder Valley Number 1 was located out in the fields just south of Erie Parkway and west of Weld County Road 5.

**State Mine.** 1913–16. The State Mine was located out in the fields just south of Erie Parkway and west of Weld County Road 5.

## Farther East of Old Town

**Clayton Mine.** 1920–42. The Clayton Mine was located in the quadrangle formed by Weld County Road 5 on the west, Weld County Road 7 on the east, Weld County Road 10 on the north and Erie Parkway on the south. This mine was slightly northwest of center and includes the grounds of the new Erie High School.

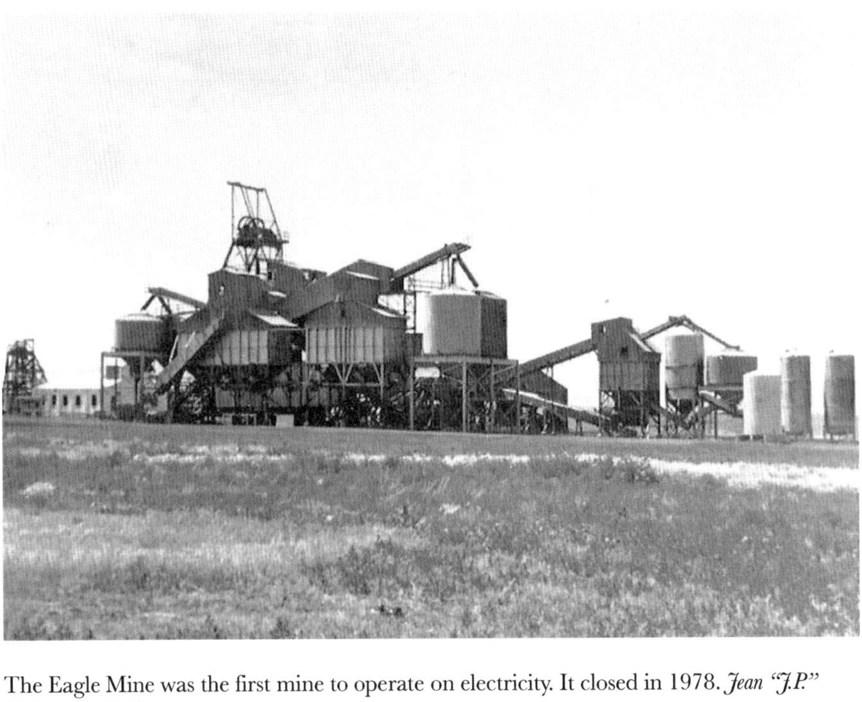

The Eagle Mine was the first mine to operate on electricity. It closed in 1978. *Jean "J.P." Holloway collection.*

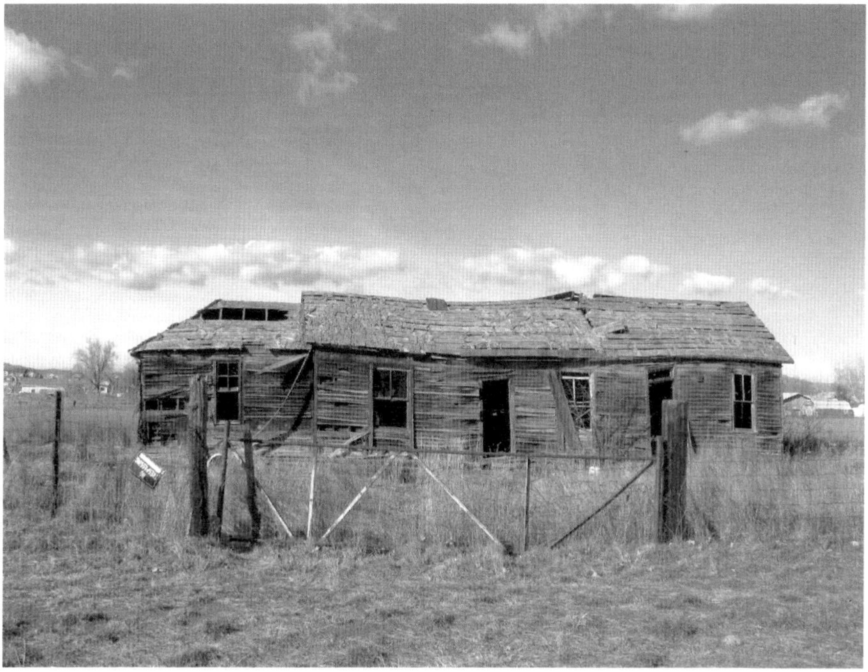

This miner's cabin was moved from the Puritan Mine to the Castro property on High Street. *Author's collection.*

**Eagle Mine.** 1939–78. The Eagle Mine was located near Interstate 25 and Erie Parkway. You can still see what is left of the water tower and several concrete blocks. More evidence is visible farther from the road.

**Imperial Mine.** 1927–72. The Imperial Mine was located just north of Weld County Road 10 and east of County Road 7.

**Morrison.** 1930–53. The Morrison Mine was located north of Weld County Road 10 and west of County Road 7. You can gain access to the mine area by following a small road off County Road 7.

**Reliance.** 1903–1907. The Reliance Mine was located east of Weld County Road 3 and south of Rue de Trust, about where Sylvia Lane would go through if it extended south.

**McKissick.** The exact dates for this mine are unknown. The McKissick Mine was located just south of the Reliance Mine, north of Weld County Road 10, east of County Road 3 and west of the railroad tracks.

**Lincoln Mine (new).** 1948–79. The new Lincoln Mine was located east of Interstate 25 in the Dacono area.

## Grandview Subdivision

**Unnamed Shaft.** The exact dates for this mine are unknown. This mine was located on Monares Lane, just south of Erie Parkway and east of Montgomery Drive.

## Near Erie Cemetery

**Northwestern.** 1892–1907. The Northwestern Mine was located just off Weld County Line Road 3, north of Erie Parkway, on an unnamed road south of Erie Cemetery.

**Boulder Valley (old).** 1883–85 and 1890–93. The old Boulder Valley Mine was located across Coal Creek from Old Town Erie, approximately between Cheesman and Wells Streets, just west of Erie Cemetery, off an unnamed road.

**Unnamed mine.** The exact dates for this mine are unknown. This mine was located west of Weld County Road 3, north of Evans Street and northeast of Cottonwood Extension Ditch, near Harvey Davidson Road.

**Unnamed shaft.** The exact dates for this shaft are unknown. This shaft was located east of Weld County Road 3, Weld County Road 1½ and the Northridge subdivision and north of County Road 10.

The first producing oil well in the State of Colorado was started in 1862, just north of Canon City. In 1928, oil and natural gas were piped into Colorado by Sinclair and Standard Oil. This decreased the need and desire for coal products. The Columbine Mine closed in 1946. From 1920 to 1946, it had produced over 7.3 million tons of coal. The Public Service Company of Colorado formed in 1928 to manage the natural gas business.

In 1979, the Lincoln Mine was closed due to a fire, bringing an end to mining in Colorado's northern coal fields. Overall, about 107 million tons of coal were removed from the mines. One estimate is that Colorado contains about 60 billion tons of coal, only 1 percent of which has been mined. Geologists believe some 434 billion tons of coal may be found as deep as six thousand feet below the surface of the earth, but it is highly unlikely that it will be mined in the near future. Coal production peaked in 1918, but mining is still active in other sections of the state, primarily on the Western Slope.

Erie began to deteriorate once mining slowed down. Businesses left the area, and trees disappeared along Briggs Street. Longtime Erie resident

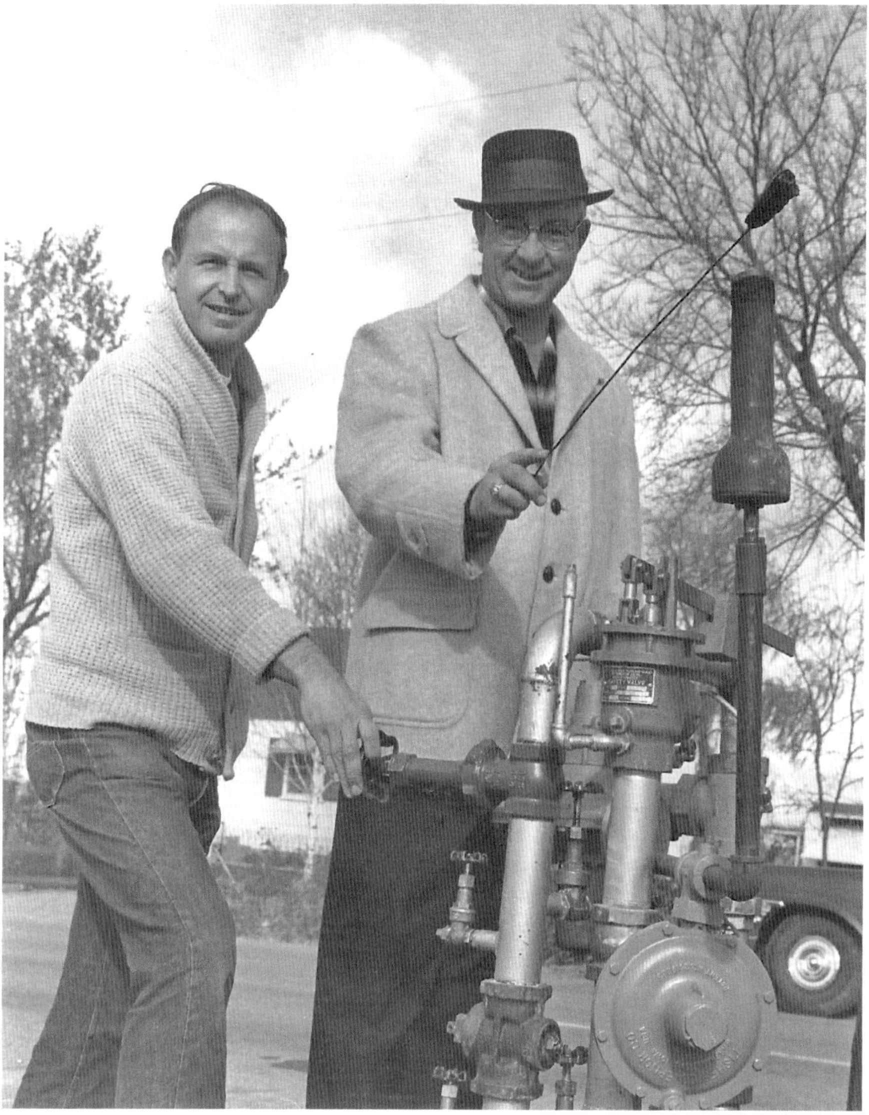

Trustee Joseph Jordan turns a valve while Mayor Sheldon Allan applies a torch to launch Erie's new natural gas system in 1962. *Carolyn Allan McCarthy collection.*

Mike Muniz recalls driving into Erie in the 1990s to find dusty streets, five bars and five churches. The schools and lodges still served the community.

By 1900, many neighborhoods began to be built over abandoned mine shafts in the Northern Coal Field, with some people experiencing subsidence, or caving in of the shafts on their properties. Some mines were left with supporting beams; in others, the beams were removed. Some subsidence will occur over a period of time, while some can take place overnight. In 1935, some of Erie's streets began to cave in because the mines were too close to the surface.

In October 2007, the Baseline Mine subsided, causing part of Bonanza Drive between Piper and Skylane to dip. It took road crews about two weeks to make it drivable again. This has become a controversial section of Bonanza, some residents wanting it to be extended to connect to Vista Parkway. In 2015, the Superior Mine subsided, causing a large hole on Cleveland Street in Lafayette.

According to a 1973 study by the United States Geological Survey and the Department of the Interior, the town of Erie does not have an undermining problem—probably because the mines were operated outside of Old Town Erie. The United States Geological Survey map shows most of the mines in the immediate Erie area had a covering of between fifty and one hundred feet. Farther away from Erie, the mines were considerably deeper.

# 8
# TRAINS AND PLANES

The best way to get to Erie before it was served by the railroads was by horseback or stagecoach. Families had to travel long distances to visit, go to school or attend church. In the 1860s, stagecoaches delivered freight, mail and passengers to communities along the Front Range. One stagecoach line ran from Fort St. Vrain (now Platteville) to Erie. Fort St. Vrain was the site of Weld County's first post office and courthouse. The Overland State Route connected Denver to Laramie, Wyoming, basically along where U.S. 287 is today.

When coal was discovered in Erie in 1866, George Gilson and others argued that without railroad service, mining was useless. Coal was needed for heating homes, powering ore smelters, running railroad engines and fueling industrial factories and power plants. It was important to be able to transport coal outside of Erie, but horse and wagon was not efficient. Many of the larger mines were owned and operated by the railroads and their subsidiaries, so it was only logical that railroad spurs be built into mining communities. The railroads leased some land to other mining companies and eastern financiers. Eventually, the railroad was less involved in mining, but very few mines were owned by local companies.

In 1867, the **Union Pacific Railroad** (UPRR) planned its transcontinental route through Cheyenne, Wyoming, to avoid the more mountainous terrain in Colorado. Feeling the pressure to connect to the Union Pacific, Colorado territorial governor John Evans and Denver banker David H. Moffat proposed a 106-mile link between Denver and Cheyenne.

Residents of Denver and Arapahoe County voted to support this project through bond issues that approximated $800,000. On November 19, 1867, the Denver-Pacific Railway (also known as the Denver, Utah and Pacific) incorporated in Colorado Territory. The railway leased and operated on this property, which eventually became part of the UPRR in 1879.

In 1870, coal mining was the driving economic force in Colorado. The Denver-Pacific Railway built a spur from Hughes (now Brighton) to Erie, which it named the Denver and Boulder Valley Railway. The spur was completed in 1871. The railroad needed Erie's coal to fuel its engines, and Denver wanted fuel for its power plant and homes; this new spur made it easier to haul coal in greater quantities. By 1871, most coal was hauled by railroad coal cars rather than by horses and wagons. However, there were still no services to Longmont, Boulder or Valmont.

According to Erie's original plat of 1871, the railroad tracks ran near Coal Creek's eastern bank. A stone depot was built on the east side of the creek by the UPRR. It also served as a telegraph office and mail stop and was operated by C.F. Wallace. Passenger fare from Erie to Denver was $2.25. The tracks crossed Coal Creek and headed west, south of Perry Street.

In 1873, train service extended from Erie through Canfield and on to Boulder. The UPRR stopped serving Erie in 1879 because problems between the railroad and mines erupted. Union Pacific pulled out of Erie and temporarily moved its operations to Hanna, Wyoming.

The railroad wanted to lure more people to the Erie area; in 1906, it published a brochure that described Erie as a metropolis. When people moved to Erie, they were surprised and probably disappointed to find a town with dirt streets and few stores. Erie had grown a little; it had a newspaper and access to a power plant. The Kuner-Empson pickle salting station provided jobs for many residents and was located by the tracks. Telephone service had been installed in town.

The **Denver and Boulder Valley Railway** stopped at Erie but did not go on to Longmont. Passengers and freight were transferred at Erie and taken to Longmont by horse-drawn wagons. Stage stop owners George Allen, Bill Dickens, Eben White and the Banner brothers were among those who provided that transfer service.

In 1878, John Thompson and others incorporated a new railroad line. Officially the Denver, Longmont and Northwestern Line, it was known almost immediately as **Longmont's Baby Railroad**. Isaac Canfield and William O. Wise were listed among the eleven names on the incorporation papers. It was designed as 8.2 miles of narrow-gauge track between

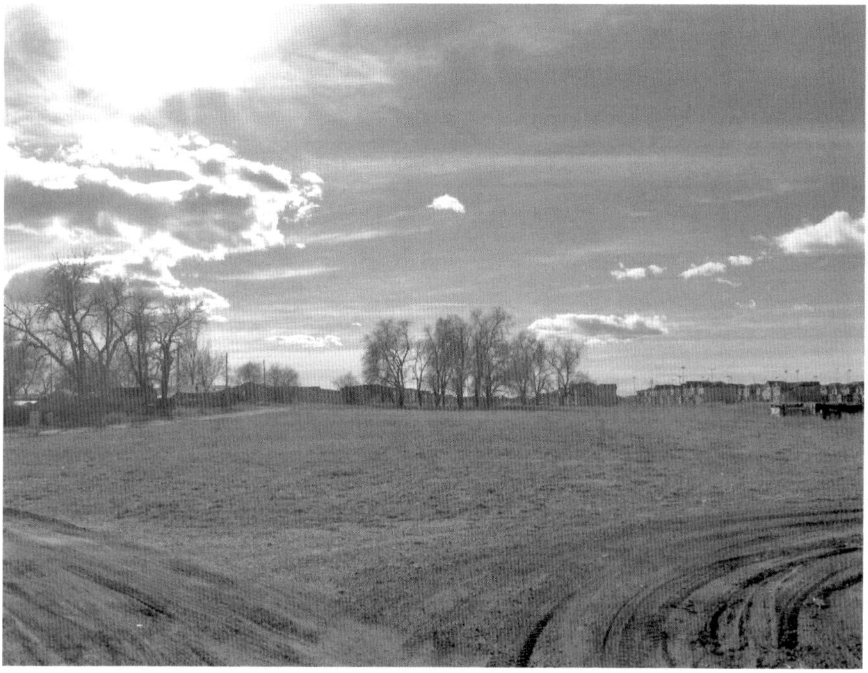

The "thirteen trees" at the south end of High Street in Old Town mark where the UPRR and Burlington Railroad tracks crossed and the locations of the UPRR depot and the Kuner-Empson pickle salting factory. *Author's collection.*

Longmont and Canfield, running along what today is 119th Street. In 1881, it was up and running in the area. It hauled coal and passengers, connecting Erie to Longmont. Canfield and Erie flourished with this line. Shortly after opening, it hauled coal from Canfield's Rob Roy and Star Mines to Mitchell at a cheaper rate than the other railroads. At Mitchell, southwest of Erie, it connected with the Denver, Utah and Pacific Railroad.

Jay Gould's UPRR retaliated with lower rates, and eventually, the Baby Railroad was closed in 1889; it was leased to a larger railroad—the **Burlington and Missouri Railroad Company**—a subsidiary of the Chicago, Burlington and Quincy Railroad. Earlier, the Colorado Southern Railroad had taken over the right-of-way and converted the narrow-gauge tracks to standard tracks. This made it possible for the Burlington, as it was known, to be rerouted away from Canfield to the eastern edge of Erie Heights on High Street, where it stopped on its way from Denver to Lyons and back. The Union Pacific had built a depot on the south end of Pierce Street. It was later moved to High Street to serve both the Union Pacific and the Burlington where the two tracks joined. The Burlington operated in Erie until 1957. Sue Larson remembers the train passing twice a day behind her grandmother's house on Holbrook. It was hauling sugar beets to Longmont. Around 1990, the tracks of the Burlington were removed from Erie, but some of the raised beds still remain.

In the 1910s and '20s, the **McKeen Motor Car**, a new type of self-propelled motorcar, ran along the UPRR tracks. It was the brainchild of William McKeen of the UPRR. Manufactured in Omaha, Nebraska, this six-cylinder, gas-powered, internal combustion engine car could be operated by one person who served as both engineer and conductor. It was nicknamed "the wind splitter" because its streamlined shape reduced the resistance faced by other trains. Diesel-powered streamlined trains were not introduced until the 1930s. The McKeen hauled passengers from Denver to Boulder, with stops in Brighton and Erie. It is sometimes erroneously called "the Galloping Goose," but the Goose was a product of the Rio Grande Southern Railway and built in Ridgway, Colorado.

In 1938, a Taylor Cub airplane, also known as a J2, transported the town of Erie's first airmail delivery. The plane had thirty-seven horsepower and no brakes—just a tailskid. According to historians, the pilot may have picked a dusty road or a flat spot near Coal Creek to land. Erie had no airport yet.

Today, **Erie Municipal Airport** is located on Colorado Highway 7 near East County Line Road. It is also known as the Tri-County Airport. In 1977, the Town of Erie annexed the 960-acre parcel from Tom Pierce and his business partners to develop the Erie Airpark. The area would include a private airport and a number of commercial and residential lots. Eventually, the developers realized by opening the airport to the public, they could gather financial support for a fixed-base operation (FBO), including the planned maintenance shop, a fuel station and a flight school. These were established by 1978.

In 1977–78, developers started to build homes at the airpark. Because of high interest rates and a shortage of water for the airpark, the project was stopped. In 1980, Pierce bought a Convair 990 airplane and had it transported from Stapleton Airport to Erie Airport. The plane was placed on pillars so developers could build a restaurant underneath that actually extended up into its fuselage. The owners leased space to restaurateurs; it was called B.J. Strawberries at the Convair. The High Plains Library District has photographs of Erie High School students in prom attire at the restaurant in the 1980s. In 1996, the deteriorating airplane was removed. The structure beneath was remodeled and is still used today as an office building.

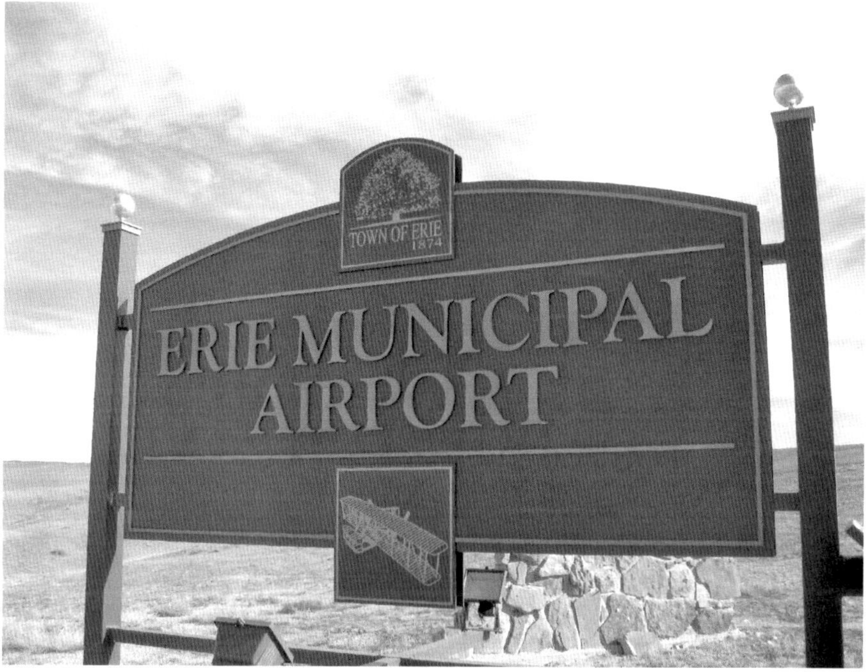


The Erie Municipal Airport, also known as the Erie Tri-County Municipal Airport, is located along Highway 7. *Author's collection.*

In 1981, Pierce died in a plane crash. His wife, Carolee, and son, Tom Jr., took over ownership and management of the airport. Also in 1981, the airport sponsored an air show featuring various acts and exhibits. A large crowd attended, giving the airport more recognition.

In 1987, interest rates improved, a water treatment plant was developed and building on the residential subdivision resumed. Today, over 140 homes exist at the airpark, with room to build more.

Airpark residents had paid into an airport maintenance fund since 1979, but it did not cover the amount necessary to maintain the blacktop runways. Tom Jr. worked with the Federal Aviation Administration (FAA) to make Erie Airport a reliever airport for Jefferson County Airport (now Rocky Mountain Metro Airport). This qualified the airport for an FAA grant and funds from the Colorado Division of Transportation. Resurfacing began immediately, but Pierce was unable to pay the matching funds; the airport fell into bankruptcy in 1991.

In 1994, the Town of Erie purchased the airport from the bankruptcy court. A manager was hired to run the FBO and pay rent to the town. In four years, Erie paid off the bankruptcy in full, and the town gave the airport a new name—Erie Municipal/Tri-County Airport.

A few organizations have formed around the Erie Airport: Erie Airpark Homeowners' Association (HOA), Friends of Erie Airport (FOEA) and the Erie Airport Economical Development Committee (EAEDC). These organizations work as liaisons among the airpark residents, existing and potential airport businesses and the Town of Erie. Over thirty businesses are located at the airport, generating about 165 jobs and millions of dollars of income. The Colorado Department of Transportation's Division of Aeronautics published that, in 1998, Erie Airport had a positive economic impact on the community. Vector Air, the airport's FBO, sponsors pilot safety seminars throughout the year.

The **Spirit of Flight Museum** is a recent addition to the airport, with Gordon Page organizing the project and bringing in artifacts and memorabilia from a nearby warehouse. This public museum charges five dollars per day or a twenty-dollar annual membership. It is open on specified

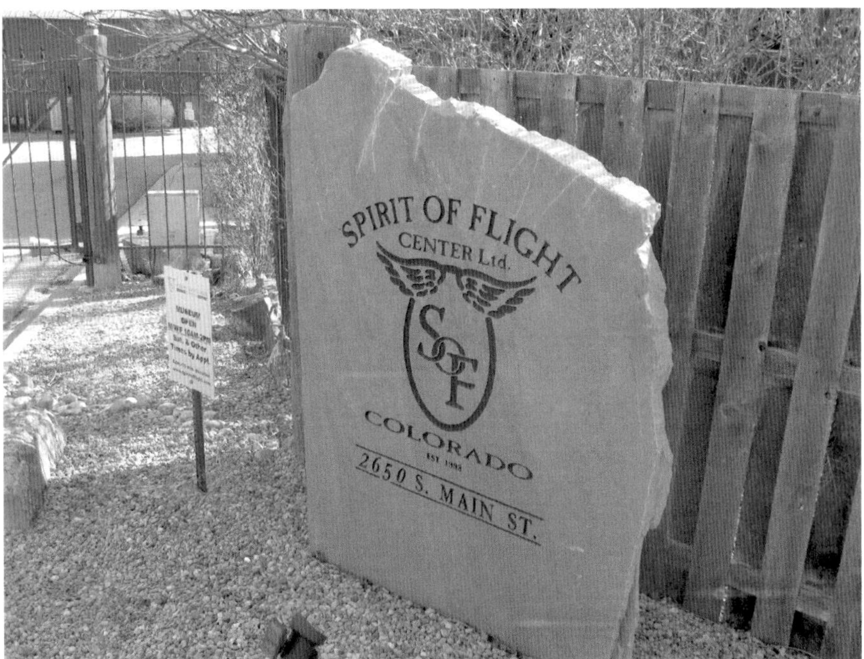


Erie's Spirit of Flight Museum is said to rival the Smithsonian Institute in terms of the quality of the relics displayed. It is located at 2650 South Main Street near the airport. *Author's collection.*

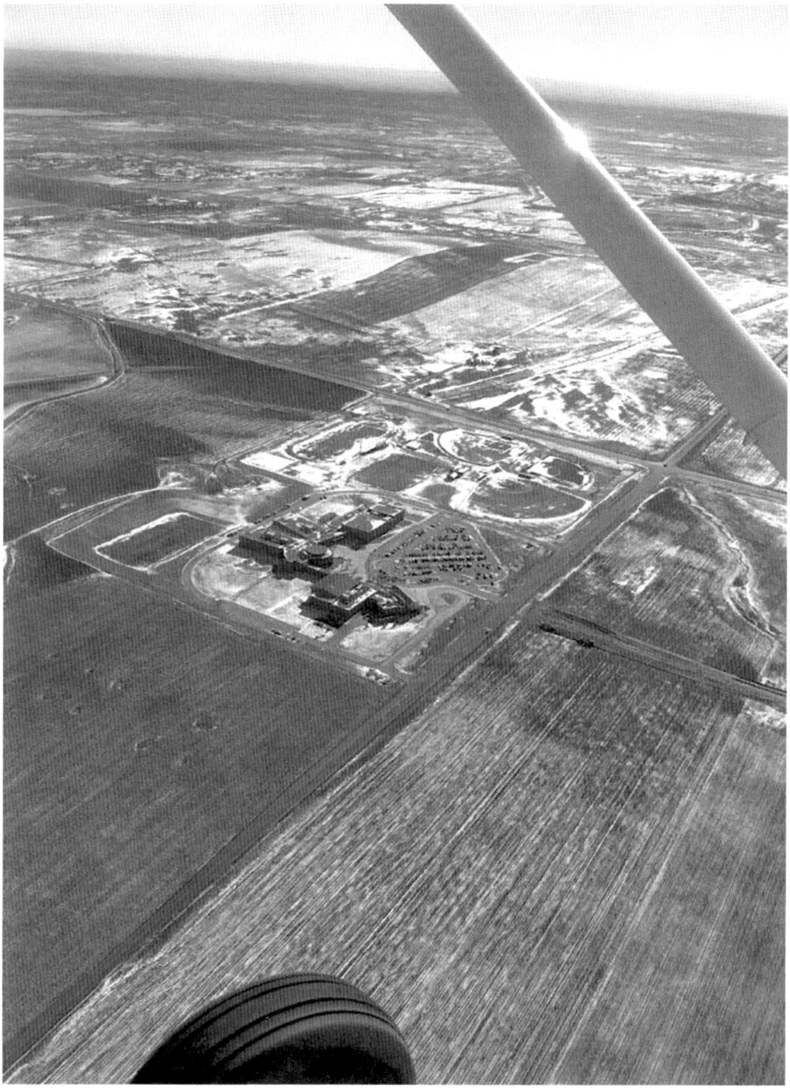

This aerial view shows Erie High School and what is left of the track of the Mile High Speedway running from the football field in a southwesterly direction. *Crystal Thornton collection.*

days. The museum is dedicated to displaying aircraft memorabilia, educating the public on the history of aviation and honoring military veterans. It holds an annual open house and special events during the year. Page calculates that over twenty-five thousand people from all over the word have visited the collection, which has items that rival those in the Smithsonian Institute.

Every year, on the third weekend in May, Erie holds its Town Fair Balloon Festival. The airport used to be the launch site for the balloons until the Colorado National Golf Course became a better venue with more space and nicer views of the Rocky Mountains. During past fairs, the airport's HOA and FOEA have provided pancake breakfasts and lunches as well as booths with displays. Some of Erie's pilots participate in the Young Eagle Rides program, giving young people a chance to fly in small airplanes. In August 2010, the airport held its own air fair.

Another neighborhood northeast of Old Town off Rue de Trust is also occupied by airplane enthusiasts, and it has several runways and hangars that are used by its residents.

# 9
# FIGHTS FOR RIGHTS

As was common in many industries in the late 1800s and early 1900s, disputes occurred between labor and management at the coal mines. Many miners, or colliers, arrived in Colorado with strong views about unionism and the rights of labor. In Colorado, there was no guarantee of a fair wage. Miners sometimes had to extract coal from the mines in water up to their waists in dangerous working environments. These conditions were labeled "industrial slavery," with miners working six days a week and twelve to fourteen hours a day. Many mining companies were able to keep their companies from bankruptcy and resurrect them by keeping their payrolls low, making miners provide their own equipment and by paying them in scrip. Power and wealth were concentrated within a small segment of society.

Many mines were governed by absentee owners who made the rules and hired mine superintendents to enforce them. To add to the exasperation of the miners, 80 percent of the coal mined in Colorado was used to fuel railroad locomotives, also owned by out-of-state companies. Railroad officials also bought up many of the mines in the area.

Mine officials and government were opposed to miners forming unions. Management referred to union advocates as troublemakers, socialists and communists—red herrings to keep people from focusing on the miners' primary goals of fair working conditions and a voice in management.

In February 1871, miners objected to how weigh bosses were screening ore. Miners believed bosses were recording the weight of the coal extracted each day as lighter than what the miners had produced, leading to lower wages.

To address this skepticism, mine operators implemented a new practice, having miners screen their own coal. However, miners were only paid for the coal extracted. Screening took valuable time away from production. Again, miners believed they were being abused. Consequently, they asked for fifty cents more per day; when this request was refused, miners struck. Management hired scab workers in their place. Eventually, regular miners were allowed back at $3.00 per day rather than the $1.50 per ton of coal produced. Since a typical day's output was two tons, this was temporarily satisfactory to the miners.

Around 1875, miners struck again. A mine owner in Canfield lowered wages to one dollar per ton, and workers immediately confronted the owner. Miners at the Baker Mine south of Erie wanted to join them but were fired as the company hired scabs. Governor Edward McCook forced the company to settle the differences. This negotiation was recognized as a significant move toward unionization. Miners not wanting to strike any longer formed the Cooperative Mining Company in December 1875 and obtained the mineral rights to the Mitchell Mine.

In 1877, someone set off explosives at the Boulder Valley Mine. Mine superintendent Joseph Bates had announced a wage cut to seventy-five cents per ton. About that time, the Knights of Labor gained support for unionization in Colorado and intervened. Bates ordered the union representatives off the property, leading to a peaceful demonstration by the miners. Judge Arthur N. Elliott of Weld County Court authorized the local sheriff to arrest the union representatives. Tensions continued to mount in the Northern Colorado Coal Field.

Children sometimes worked in the mines. Boys were allowed to work at age fourteen; however, some workers were as young as five years old. The age limit was then raised to sixteen for both boys and girls, and they needed to prove they could read and write. Mine operators were notorious for overlooking these restrictions. In 1897, Charlie Liley quit school after the third grade at ten years of age. His father told him he would have to get a job. He became a "trapper" at the Mitchell Mine. His job was to open and close big wooden tram doors to deflect the air while mules were led into the mine to pull the coal cars.

Miners typically rose before sunrise in the winter to get to work by 7:00 a.m. After their shifts, they walked home in the dark. In the winter, they often had frost on their skin and clothing from their own sweat and from the water on the floors of the mines. If miners lived in homes with a tub, they would bathe there. Those living in boardinghouses had to clean up in the washhouses in the backyards before having dinner. They also had to wash their own laundry.

Miners were independent contractors, meaning they bought their own equipment. If they lived in a mine camp, they paid rent on company-owned property. They paid their own medical expenses, often to company doctors. They wanted to be paid in cash or gold, rather than in company scrip that could only be used at the companies' stores, which typically charged higher prices than the stores in the towns. The value of the purchases was deducted from the miners' pay. Miners often found themselves in debt to the company because of this arrangement. They could never earn enough to meet their needs, but the company was happy to let them continue to purchase items on credit.

Miners wanted to work eight-hour days instead of the twelve- to fourteen-hour days companies required of them. Miners believed eight hours was enough time to work underground. In 1889, the Colorado General Assembly passed an eight-hour workday law. The Supreme Court declared it unconstitutional because no legal body should dictate the number of daily hours a laboring man might work.

In 1873, an international recession was felt in the Erie area; it lasted until 1879. This panic caused railroad officials to pressure mine operators to lower their coal prices, which caused the operators to threaten to lower miners' wages from \$1.25 per ton to \$1.00 per ton. Miners organized under the Miners National Association (MNA) and discouraged Boulder Valley Mine operators from cutting their wages. In 1875, mine operators cut miners' wages by \$0.25 per ton; this resulted in the miners vandalizing chutes, boardinghouses and other mine property.

During the summer months, the mines were closed because of the lack of need for coal and because the soft lignite coal was too unstable to store for any length of time. Miners often worked in the sugar beet fields or harvested other crops for the local farmers. Others worked in the gold and silver mines in the Rocky Mountains. Wives took in laundry or did housework for others, while children scavenged the railroad tracks for fallen pieces of coal to be used for the kitchen stove. If someone was lucky, like Joe Fletcher, he might get a job working for the railroad. Mike Brennan's sons worked at the pickle-salting station and delivered papers for the *Denver Post*.

In 1876, 100,000 tons of coal were mined in Colorado. In 1900, this climbed to 525,000 tons. Colorado was the largest coal-producing area west of the Mississippi River. Over 11,000 miners worked in the state, with about 1,650 working in the Northern Coal Field. With this boom, large conglomerates easily recruited unskilled laborers and immigrants to relocate and take non-union jobs during strikes. Company representatives filled their heads with promises about the new land and employment opportunities. People poured in from France, Belgium, Poland, Italy, Croatia, Greece and Russia. They were not told that they were coming to America to replace existing miners. Once they began working at the mines, they became easy prey for union representatives, who would intimidate them and warn them of the dangers involved in taking jobs away from the more experienced miners. Regular miners feared this competition. This led to distrust among all factions of the mining industry.

Chinese and African American workers often went to the Southern Colorado Mine Fields, where camps were set up for them. This form of paternalism existed at the mines in both coal fields. Owners provided housing and company stores for the miners in an attempt to exercise more control over them and keep the unions out.

Because miners' interests were in jeopardy, workers' rights groups emerged in Colorado, including the Knights of Labor, the Western Federation of Miners, the United Mine Workers Association, the Miners National Association and the Industrial Workers of the World. Because unions were not welcome, miners and union organizers needed to be secretive about their affiliations.

By 1878, the Miners National Association ceased to exist. About sixty Erie miners formed Local 711 of the Knights of Labor—the first national union in Colorado. Representatives visited the Colorado coal fields to persuade miners to join the union and focus on improving working conditions. Unions faced a challenge; they needed to convince mine owners they truly represented the voice of labor and that collective bargaining should be allowed to improve working conditions. Mine owners insisted that they contracted with individual workers and that's how workers' rights would be determined.

Colorado strikes tended to start in the Northern Coal Field and spread southward. In 1885, the Knights of Labor called for the first widespread strike in the Northern Coal Fields, and the miners voted to strike. In 1886, the Coal Miners Federation of Northern Colorado (a Louisville organization) worked to improve working conditions for local miners, including no blacklisting of miners and better child labor laws.

In 1887, miners earned about twenty-one dollars for a sixty-hour workweek; they worked ten hours per day. Young boys earned six dollars per week. Mule drivers earned eighteen dollars. Firemen earned fifteen dollars.

Weigh bosses and engineers earned eighteen dollars for a seventy-hour week. Foremen worked extended hours and earned between twenty-one and twenty-four dollars per week. Two dollars were deducted from each miner's wages for medical care. A doctor was always on the mine payroll and tended to miners' illnesses and injuries. In some cases, if a miner needed special attention, additional payment for the doctor was withdrawn from his wages. Some unethical doctors sided with the mine owners and deemed miners ready to work even though they had an illness or injury.

Also in 1887, unions convinced the Bureau of Labor Statistics to meet their responsibility to enforce labor legislation. An anti-blacklist law was established, disallowing mine officials to refuse employment to miners who may have expressed concerns about working conditions. In 1887 and 1889, two statutes were enacted permitting labor groups to exist without being considered conspiracies, but collective bargaining was not yet a practice at this point. By then, miner-chosen weigh bosses were mandatory by Colorado law, but most mine operators disregarded the law.

In 1888, some workers from the Knights of Labor formed the Progressive Union of Miners and Mine Laborers. In 1889, local members of the UMWA attracted members of the Knights of Labor. UMWA membership quickly went from sixty to ninety. Other miners' groups followed this movement.

Miners claimed management was not measuring the gas levels in the mines. Carbon monoxide and methane were problematic. Explosions and cave-ins were common. Black lung disease was responsible for the death of many miners. The average life expectancy of miners was under forty-five years of age. Colorado's mine inspector reported that mines had poor ventilation and drainage. Mine rooms had weak walls and often crumbled into adjacent rooms, and timber structures were weak from age. Workers' lamps and blasting powder often exploded unexpectedly, leading to injuries and death. Mine owners and operators held miners culpable for any accidents, taking no responsibility to ensure that safety standards were established and upheld.

In 1890, the Rocky Mountain Fuel Company began purchasing mines in the Northern Colorado Coal Field.

In 1893, the American Protective Association (APA) grew in Colorado to a membership of three thousand. Association members were primarily Protestant miners who didn't want scabs taking their jobs. They were anti-Catholic. Colorado's APA membership eventually reached about ten thousand. Many members eventually joined the Ku Klux Klan.

In 1898, mine owners persuaded the United States Legislature to ban employees from picketing or hindering scabs or strikebreakers from working.

In 1900, the United Mine Workers Association established District 15, which encompassed Colorado, New Mexico and Utah. At first, this movement was kept quiet, but soon twenty locals had been organized with a total of 2,470 members. Miners in Erie, Lafayette, Louisville and Marshall formed the critical mass for the association's reentry into Colorado mines. The organization's president was John Mitchell.

In 1901, an act was passed that gave miners a semi-monthly paycheck. In 1902, an eight-hour day constitutional amendment was passed.

In 1905, the Industrial Workers of the World (IWW or "Wobblies") formed. The IWW was considered the most radical and violent union. Because of its reputation, its headquarters in Pueblo was raided, and an ordinance was enacted in Denver forbidding anyone from speaking out in a manner that might incite rebellion.

During Governor John F. Shafroth's administration from 1909 to 1913, he passed a coal mine inspection law.

In 1910, labor strikes closed mines in the Northern Colorado Coal Fields for four years. In 1911, wages were raised voluntarily to three dollars per day. In 1912, miners were allowed to work an eight-hour day. But miners continued to strike. In 1911, the Erie Commercial Club pushed for an increase in farming to boost Erie's economy. Erie's population was dropping due to inactivity at the mines.

According to Joe "Cotton" Fletcher, miners later received $3.75 per hour for an eight-hour day until the beginning of World War I. After the war, hourly wages rose again but dropped to $5.00 during the Great Depression. Many mines were forced to close during the Depression. Some businesses extended credit to the unfortunate miners' families, but the nation's slow recovery eventually forced many of these businesses out of operation.

In 1911, the Mitchell Coal Company and the Economic Coal Company joined to form the American Fuel Company; this new company quickly signed a contract with the United Mine Workers of America.

In 1914, the Colorado mining industry suffered the worst tragedy in its history at the Ludlow Mine in southern Colorado, north of Trinidad. It was opened in 1883 by the Colorado Fuel Company, the largest coal mining company in Colorado. John Osgood was a major player in the development of Ludlow. He believed that owners and operators needed to be firm to show their power over mine workers, but he initiated a program to improve the company's image and keep up the morale of mine workers. He formed a mining camp called Redstone, which provided better sanitary conditions and contained housing, a movie theater, a school, a social club, a saloon, a bathhouse, a laundry facility, a barn for livestock and a garden for miners to grow their own vegetables. He believed providing these amenities for miners might keep them from joining unions.

Osgood eventually lost control over the operations in 1892 when he sold the company to John D. Rockefeller and Jay Gould. They changed its name to Colorado Fuel and Iron, with headquarters in Pueblo. Rockefeller purchased many mines and the Colorado and Wyoming Railway. Rockefeller, Gould and other owners were opposed to Osgood's "welfare capitalism" and immediately moved miners and their families into union-provided tents. The situation at the mine worsened. Owners ignored warnings from Ethelbert Steward of the United States Department of Labor, deputy labor commissioner Edwin Brake and others about the trouble brewing at Ludlow.

Miners demanded recognition of their union. They also wanted a 10 percent increase in wages, stricter enforcement of Colorado's eight-hour work day law, health and safety regulations, the right to select their own living quarters, reasonable places to eat and medical care. In 1913, at the age of seventy, Mary Harris (aka Mother Jones) visited Ludlow to speak to the miners about their rights. Ed Doyle, secretary of the UMWA District 15, was also present to encourage workers to stand up for their rights but to avoid causing trouble. The vice president of the UMWA said the working conditions in Colorado mines fields were the worst in the nation. Southern Colorado Coal Field miners called a strike. This raised the number of miners on strike in Colorado to between eight thousand and twelve thousand.

John R. Lawson was a leader of the UMWA District 15. After he quit school in 1879, he began working in the coal mines at the age of nine. Originally with the Knights of Labor, he helped organize the UMWA in 1890 and became a member of its board of directors in 1907. He went to Trinidad to organize the union there, as there were some thirty-eight miners' camps between Trinidad and Walsenberg and west to the Rocky Mountains.

In 1914, detectives (hired guns and thugs) from the Baldwin-Felt Agency in West Virginia arrived at Ludlow to control miners and their families in their tent city. They lectured new miners and scabs who were not savvy about the oppression they faced as miners. They oversaw the mines and discouraged strikes. The detectives were heavily armed and even had an armored vehicle nicknamed the "Death Special." They installed six machine guns in strategic positions around the mine area and camp.

The UMWA issued a statewide call to arms via newspapers and telephones. Miners who had been evicted from company housing for striking were assisted by the association. Miners and their families from Erie and other Northern Coal Field areas donated wages and supplies and struck in sympathy of miners from the Southern Field. They sent fifty miners, along with another two hundred from Lafayette and Louisville, to Ludlow.

Governor Elias Ammons summoned the Colorado militia to keep order at the mine and break the strike. Two militia members with anti-miner attitudes and a lack of discipline were Adjutant General John Chase and Lieutenant Karl Linderfeldt. Chase commanded nearly one thousand troops at Ludlow, whose duties included protecting strikebreaking miners and harassing striking miners and their families. Chase regularly blamed miners for any incidents in the area and put them in jail without allowing them to see an attorney or have visitors. Whenever miners claimed their rights were being denied, Chase responded that they had no rights under martial law. Under Chase's watch, miners' tents were often searched for guns and alcohol by militiamen who helped themselves to money, watches, knives and anything else they considered valuable.

On April 20, a violent battle ensued, resulting in nineteen deaths: one soldier, two miners, three union leaders, two women and eleven children. The thirteen bodies of the women and children were found—burned and asphyxiated—after the tent city had been torched and destroyed by the detectives. The precedent had been set: well-armed outside enforcers, including federal troops, could deny miners their rights.

Lieutenant Linderfeldt was a hated and feared man. He fought for the United States in the Philippines, hired on as a mercenary in Mexico and served as a mine guard and militia member at Cripple Creek. He is credited with having led the torchbearers who razed Ludlow's tent city. He also crushed the head of miners' spokesman Louis Tikas with his rifle butt and then shot Tikas three times in the back, murdering him as he struggled to flee. Linderfeldt admitted in court that he injured Tikas, but he claimed he had rushed into tents during the fire and saved numerous children. He was acquitted of most charges and reduced in rank for striking Tikas.

Union officials wrote letters to President Woodrow Wilson about the incident. Mine operators wrote their own accounts; as expected, these accounts differed dramatically. Mine records of people who lived and died there were burned, so no official accounts are available. One positive outcome from the Ludlow incident was that industrial relations were receiving increased attention. John Lawson was arrested and convicted of the murder of a soldier. The Colorado Supreme Court eventually overturned the charge.

Eventually, Colorado Fuel and Iron granted the union a sixty-eight-cents-per-day increase and asked that the miners divorce themselves entirely from the IWW. The Southern Coal Field calmed down after this compromise, following one of the largest labor uprisings in United States history. However, the precedent was set. With the backing of the mine owners and the government, the Colorado militia had been successful in quashing the strike at the Ludlow Mine without any major repercussions.

On April 27, 1914, a gun battle broke out in Lafayette between miners and mine officials at the Hecla Mine as a reaction to the Ludlow incident. Miners from Erie joined the fracas. Zell Wheeler drove a wagonload of men to Lafayette, where they set up behind piles of straw and fired weapons at those trying to control strikers. In court, nobody identified the Erie men involved in this incident. Wheeler conveniently "lost his memory," stating he didn't remember hauling strikers. M.T. Alumbaugh testified that he had been asked originally by William Knowles to take miners to Lafayette. The outcome was that Wheeler was cited for perjury and jailed for three years. William Knowles, Arthur Spendlow, Gus Brack and Dan Griffith were four of eleven men indicted for the attempted murder of William Bucklin, who had been shot in the hand. Knowles and Griffith were found guilty, while Spendlow and Brack were acquitted.

In November, President Wilson called for federal mediation of the longest strike in Colorado history. Miners eventually declared peace with Ludlow and Northern Colorado mine owners. However, it was clear to miners and union officials that a strong link existed between government and industry leaders. Conditions improved temporarily for mine workers because the world had watched the Ludlow situation in a state of shock.

In 1914, the United States Commission on Mines and Mining reported that Colorado had good mining regulations designed to protect the miners by ensuring safety in the mines, if they were enforced properly. However, the percentage of mine-related fatalities in Colorado was larger than in any other state. The commission concluded that something was wrong with the way the coal mines were managed.

In 1916, John Lawson was elected president of District 15 of the United Mine Workers of America. He had been released from prison for his involvement at Ludlow. The union did not choose to recognize District 15. Feuding and other internal union problems resulted. In 1917, all 4,500 miners in the Northern Coal Field stayed home in support of a strike at the Columbine Mine. Again, mine owners hired scabs.

In 1917, the Colorado National Guard was formed—the first in the nation. Governor Billy Adams did not want to use the guard against miners, so he established a special military unit to deal with mines. He selected Louis

N. Scherf, a military aide to Pat Hamrock, to run this special unit. Hamrock was a principal figure in the Ludlow massacre. This new unit was dubbed unofficially as "the state law enforcement department." Scherf heavily armed his unit of twenty men. He trained them in armed riot control and ordered airplanes equipped with machines guns as he eagerly prepared for the next confrontation. He even procured a Douglas bomber and three other airplanes from New York.

In 1918, the UMWA became known as United Mine Workers. Besides dropping the end of its name, the organization also dropped any miner who carried an IWW card, as the IWW was regarded as too radical. John Lawson and Ed Doyle left the UMW to form the Independent Union of Coal Miners of America. The UMW criticized them harshly for their short-lived, divisive venture. In 1919, World War I ended. Some 450,000 miners were idle throughout the country.

In 1920, miners struck for equitable pay for other jobs they had to do that interfered with their ability to extract coal, including cleaning out slag from tunnels (narrow work), setting timbers to support mine shafts and tunnels (dead work) and laying track for the coal cars. Miners believed mine operators were still treating them unfairly. In 1922, miners struck at fourteen mines in Boulder and Weld Counties. The Colorado National Guard was called in to handle any problems.

The Columbine Mine was one of the largest coal mines in the Northern Colorado Coal Field, producing 7.5 million tons by its closure in 1946. It was owned by the Rocky Mountain Fuel Company (RMFC). It opened in 1920, and by 1927, it employed some 350 miners. Its camp, Serene, was located on a rolling hillside southeast of Erie, near the current site of the Erie Landfill west of Weld County Road 5. Its location and access cut Serene off from outsiders. It was a collection of dirty company houses surrounded by a barricade of barbed wire. It was illuminated at night by a large searchlight that was installed on the mine tipple. Serene had a post office, a church, a schoolhouse, a general store, a mess hall, a dance hall, a brothel and a casino. The mine included a main shaft, a tipple, a water tank and administrative offices. The conditions at Serene contributed to the dissatisfaction of the miners and their families.

In 1927, the IWW was the voice of Colorado miners. Adam Bell established himself as the IWW leader in Erie around 1926. The IWW organized a strike in 1927, shutting down all of the mines in Northern Colorado except the Columbine. The mine was now operating with 150 scabs, who were enticed by management to work with a fifty-cent-per-day wage increase. Throughout Colorado, 113 mines were closed, and 8,400 miners were on strike. Statewide, 13 mines remained open, employing 1,750 scabs. The IWW claimed workers had no redress of grievances through the State Industrial Commission. Miners had suffered two wage cuts in 1925 and wanted improved working conditions. The IWW intentionally scheduled the strike for the winter season. By mid-November, the weather was cold, and not enough coal could be mined to supply the Valmont Power Plant, the University of Colorado, hospitals, railroads, businesses and homes. The shortage of coal became critical.

The IWW had organized one of the best-run coal strikes in Colorado history at the Columbine Mine. Union representative Frank Jurich came out from Chicago and formed Industrial Union 210–220, the Colorado unit of the IWW. While the press attacked the presence of the IWW, neighboring communities ignored the criticism.

The miners had the permission from the RMFC controlling stockholder to strike. Josephine Roche was sympathetic to labor and workers' rights to fair wages and safe working conditions. She believed happy workers were

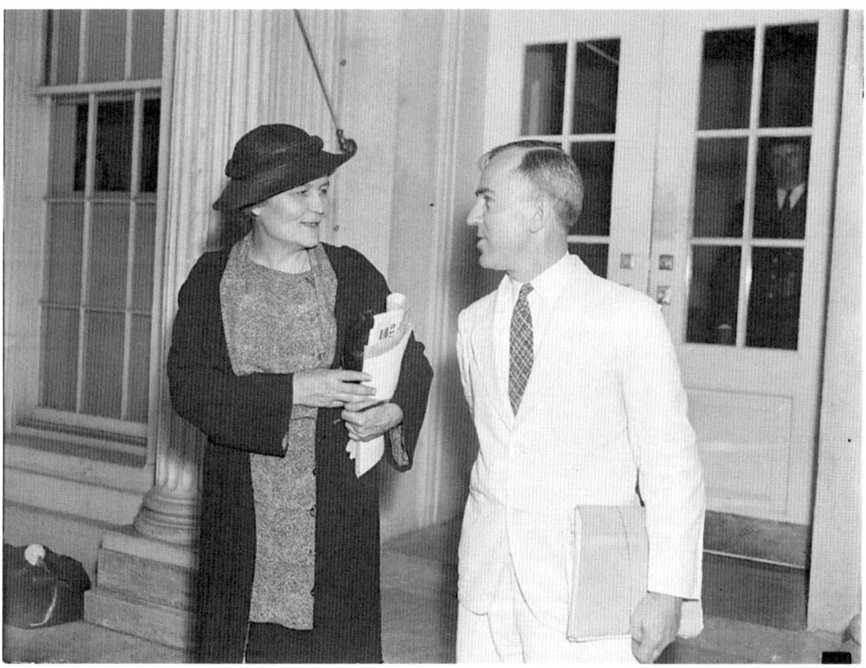

Josephine A. Roche (1886–1976), assistant secretary of the treasury, chats with John Studebaker, commissioner of education, in 1935. *Library of Congress.*

productive workers. Her social work experience led her to understand that workers' grievances might truly be legitimate complaints. Her position on workers' rights led to the horror of other RMFC owners. Many withdrew, and she bought their shares. She now controlled a multimillion-dollar enterprise in which she could experiment with her ideas. Because of the significance she eventually had on increasing wages, decreasing working hours and improving conditions in general in the mines, mine workers supported her with "Buy from Josephine" slogans on coal bags, railroad cars and in chants.

Governor Adams stated that if miners didn't like the conditions of their employment, they should quit. This was the official recommendation of Adams's Industrial Commission. Miners were quick to respond that they had the right to go to the Serene Post Office, to pick up their children from the Serene school and to demonstrate lawfully. They demonstrated at the mine every day. Groups of men, women and children picketed outside the northern camp entrance and then marched through the camp to the post office and back. Elizabeth Beranek, age forty-four, often led the picketers. She was the wife of a striking miner, Joe Beranek. She was also the mother of sixteen children—ten girls and six boys. She was expecting her seventeenth child as she demonstrated. All of her sons eventually worked in the mines. She only stood five feet, two inches tall; however, the Denver newspapers referred to her as "the Amazon" because of her no-nonsense manner.

Each day, the demonstrations became more verbally confrontational, with guards calling men cowards for letting women and children march at the front of the picket line. Wanting to keep scabs and other strikebreakers inside the camp and strikers outside, mine owners increased the number of guards at Columbine from two to nine.

On November 7, 1927, strikers and automobiles blocked the road between Lafayette and the Columbine Mine. They had the support of miners and their families from surrounding towns as far away as Marshall. On November 8, strikers repeated this tactic. Adam Bell organized a group of one thousand miners to visit several mines one morning to show spirit and gain support. He also spoke at a rally on November 13 in Denver.

On November 14, following scuffles between miners and law enforcement, Weld County sheriff Ben Robinson ordered the Columbine Mine to be closed. A machine gun was moved to the top of the water tower and another to the top of the tipple. Some historians believe the tower on the mine head might have been a spot for a third machine gun. More barbed wire and sandbags were brought to the site. Four airplanes regularly patrolled the area above Serene. At least one plane was equipped with yet another machine gun.

People from Erie and Serene frequently reported that guards were shooting from the tipple or water tower at targets in Erie. The guards became a nuisance, harassing residents of both communities and using violence to shut down union meeting halls.

The RMFC wanted to settle the strike on its own terms, but in desperation, Governor Adams summoned the National Guard and police. Louis Scherf rounded up Wobblies at Columbine and jailed them. His actions resulted in IWW workers from all over the United States converging on Erie in support of those arrested and those striking. On November 20, after five weeks of strikes and two weeks of rallies at Columbine, Adams ordered his lieutenant governor and the chairman of his Industrial Commission to Columbine. Rather than leave immediately, they drove up from Denver at dawn the next day. Typically providing coffee and doughnuts for the striking miners and trying to keep peace, Josephine Roche was in Denver on business and was not informed of the escalation of events at the mine.

Between six hundred and one thousand miners were waiting at the mine gate early on the morning of November 21. Adam Bell and Elizabeth Beranek led the group. One of the mine guards remarked that if the miners came in, they would be carried out. The presence of the militia signified the seriousness of the situation. They were dressed in plain clothes but were armed with pistols, rifles, machine guns and tear gas grenades. Scherf had also ordered hard hats from Denver. The night before, he secretly moved his entire team into Serene—not with the authorization from Governor Adams but from the encouragement of leaders at Colorado Fuel and Iron at Ludlow.

The guards taunted the miners and refused to allow them into Serene. Miners stormed the gate and continued demonstrating. Bell was dragged over the fence and beaten by guards. Beranek got through the gate and covered Bell with an American flag. Both she and Bell were struck continuously. A Mrs. Kubic is reported to have been hit in the back with a launched tear gas grenade.

Scherf fired a couple of warning shots. Taking this as a signal to engage, the militia opened fire on striking miners inside the gates. Lawrence Amicarella hit the ground when the shooting started, as did Mary Borstnick. Jessie Irwin turned away from the road to Serene and ran back to Lafayette and his car. John Eastenas, thirty-four-year-old father of six from Lafayette, was the first miner killed. He died immediately. Nick Spanudaklis, a thirty-four-year-old

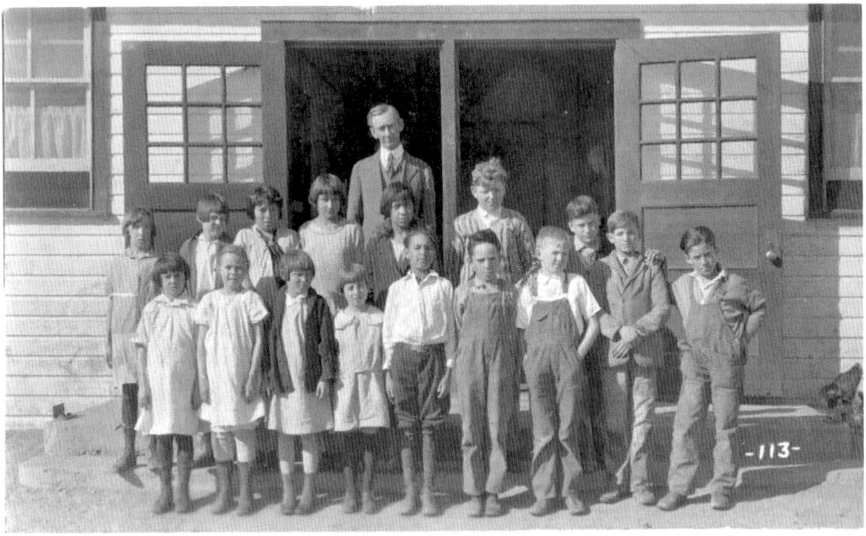


Principal Griffith and grades one through six pose for a photo at the Columbine Mine School at Serene around 1927. Henry Amicarella is in the front row on the far right. *Lafayette Public Library.*

bachelor from Lafayette and striker from the Morrison Mine, died a few hours later in the office of Dr. Clarence Bixler. Shortly after, three others died from their wounds: Jerry "Jay" Davis, thirty-four, from Frederick; Rene Jacques, twenty-six, from Louisville; and George Kivitch, twenty-five, from Erie. Mike Vidovitch, thirty-five, died a few days later in a Longmont Hospital. The toll was six dead and around sixty injured, including women who insisted on supporting their husbands. Three guards received superficial injuries. Henry Amicarella recalled watching his mother and sisters shred bed sheets to make bandages for the wounded miners. Doctors attending to the wounded reported extracting machine gun bullets from the miners. At one point, five of the victims' coffins were displayed in Frederick.

Martial law was declared in the Northern Coal Field. People stayed off of the streets in Erie. Governor Adams supported the actions of Scherf and his militia as well as those of the National Guard. He was opposed to any talk with the unions. He called additional National Guard troops into action, deploying 325 soldiers and forty-five cavalry, along with two whippet tanks. Airplanes continued to patrol the area. Guards were also posted at the Puritan, Grant and Monarch Mines. Passes were necessary to get in or out of the mines. A curfew was established. This incident is said to have led to Adams's withdrawal from politics.

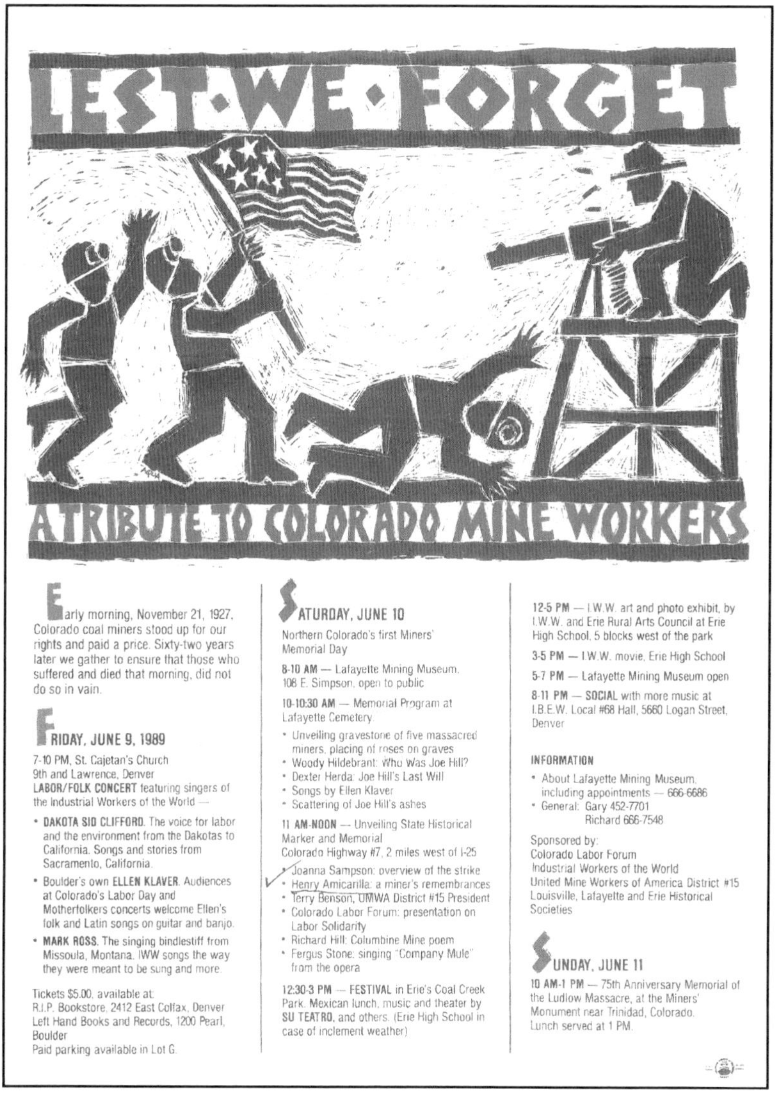


This graphic by Hal Aqua symbolizes the struggle miners faced to bring about favorable working conditions. A commemoration was held in 1989. *Marianna Amicarella collection.*

In court, the three mine guards showed up heavily bandaged to gain the sympathy of the judge. They claimed they had only shot pistols and rifles into the air to discourage the rioters. Adam Bell, Tom Harris, George Kubick and James Allender were jailed in Longmont. There was no contest, as the government continued to oppose strikes by miners.

Miners eventually went back to work after losing yet another battle. However, Josephine Roche brought an end to the strike two weeks later. She recognized the American Federation of Labor and the United Mine Workers Association as unions she would permit within the RMFC.

10

# IMPROVED WORKING CONDITIONS

In 1931, Rockefeller's Colorado Fuel and Iron Company dropped its wage rate. Josephine Roche reacted openly to this move and gained the respect and support of miners. At this point, the Colorado Industrial Commission was still clearly in favor of the mine owners and anti-union.

In 1932, the Colorado Fuel and Iron Company petitioned to lower wages again. This time, the Industrial Commission refused their request, claiming that miners already did not receive a decent living wage under the present scale. Colorado Fuel and Iron defied the Industrial Commission and lowered wages anyway, continuing its industrial slavery.

Josephine Roche was forced to lower wages at the Rocky Mountain Fuel Company (RMFC), but she also lowered the prices in her stores. She even got a loan to ensure that miners were paid when they were on strike.

Roche's RMFC was second in holdings only to Rockefeller's Colorado Fuel and Iron Company. Josephine gained 51 percent of the company's stock in 1938. She elevated Merle Vincent to president and hired former labor organizer John Lawson to serve as vice president. Thomas Costigan was appointed as general counsel. Costigan did not like trying to settle with the IWW, even though they had led the strike at the Columbine Mine and influenced the unionization efforts at the RMFC. Consequently, Josephine met with UMW officers and signed a contract on August 16, 1928, organizing three-quarters of the Northern Colorado Coal Field. Her goal was to bring order to what was regarded as the most disorganized industry in the United States. Rocky Mountain Fuel Company employees now would work an

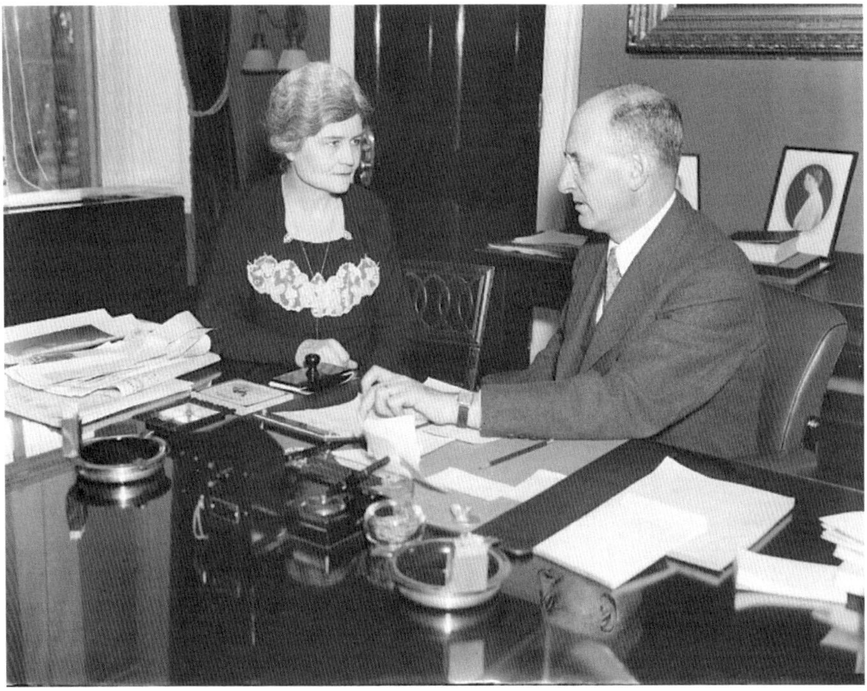

Josephine Roche meets with her boss, Secretary of the Treasury Henry Morgenthau Jr., around 1934. *Library of Congress.*

eight-hour day six days a week and earn seven dollars per hour. They would receive eight paid holidays and be paid twice a month. Each mine would employ a weigh man, and pit committees would handle grievances. Two years later, Costigan was a United States senator representing Colorado. Josephine Roche is considered the hero who paved the way for union organization of the Colorado coal mines.

Roche ran for Colorado governor as a Democrat and lost. Recognized for her leadership and talent by Franklin D. Roosevelt, Josephine went on to hold numerous government positions in Washington, D.C., and other locations. She served as the assistant secretary of the treasury and supervised the Public Health Service. She directed the National Youth Association. She was considered the second most important woman in Roosevelt's administration—second to her college friend Frances Perkins, secretary of labor.

Roosevelt signed the National Industrial Recovery Act of 1933, mandating that all United States coal companies allow their employees to unionize. From that time, miners were allowed to join the UMW. Many local mines reopened, and miners struck for better wages and working conditions. The Wagner Act of 1935 legalized collective bargaining. The National Labor Relations Board outlawed the Rockefeller plan that established company unions. Even with improved working conditions, miners continued to experience mining accidents and deaths.

Colorado's coal industry saw a rebound during World War II. However, Roche's Rocky Mountain Fuel Company was slowly dying, producing less coal each year. In 1944, the company declared bankruptcy. Cars and trucks were using oil; train use declined. Natural gas and electricity were quickly replacing coal, and transporting it became much easier with seamless, electrically welded pipelines. Some people believed the IWW strike was a fraudulent effort by oil companies to ease the introduction of gas franchises into Colorado.

Josephine returned to Washington, D.C., to be the director of the United Mine Workers pension fund. She worked there until her death in 1971.

By 1960, most coal mines were no longer producing around Erie, and they were closed. In 1967, the Washington Mine closed. In 1972, the Eagle Mine discontinued operations; part of the water tower there remains and can be seen from Erie Parkway, just west of Interstate 25. The New Lincoln Mine closed in 1979; it was east of Interstate 25, but it employed many of Erie's miners.

As the mines closed, workers went into construction and other lines of work. The Atomic Energy Commission opened Rocky Flats, and some miners went to work there.

The federal government set up a Black Lung Benefits Program in 1969. The Federal Coal Mine Health and Safety Act of 1970 gave $136 per month to miners or their widows. Although some doctors were paid by coal operators to deny any such disease, the United Mine Workers pointed out the about four thousand miners died each year from it.

The history of the coal industry in Colorado is often underplayed or forgotten. The significance of the strikes needs to be chronicled. The results led to more effective labor organization throughout the United States in all industries. Benefits that workers received today may have taken longer had the nation not paid such close attention to the Ludlow and Columbine incidents. Josephine Roche's focus on workers' rights is visible in contemporary management theories and practices that place considerable emphasis on concern for both production and people.

However, things often regress. In 1947, the United States Navy medical corps reported that cronyism between mine operators and doctors resulted in terrible service for miners; working conditions deteriorated. In 1946, only 2 out of 3,345 mines complied with the Federal Mine Safety Code.

On April 29, 2000, the Coal Miner's Memorial was dedicated in memory of Clarence Bixler, the doctor who attended to those wounded and killed at the Columbine Mine massacre. The committee responsible for organizing this project consisted of Margie and Chuck Jordan, Maxine and Bib Canaday, Don Simansky and Ray Montano. Many names are missing due to the transience of so many of the thousands of miners who came to Erie during the 113 years it was a mining town. The memorial is updated periodically when families and friends report new names. The names of the six miners killed in the massacre are followed by an asterisk. Other miners killed or injured in line of duty include Henry Betts (crushed at the Eagle Mine), Sam Clark (electrocuted at the Shamrock Mine), John Grimson (cave-in at the Lehigh Mine), George Cook (cave-in at the Lehigh Mine), Pasquale Lombardi (falling rock), Job O'Taylor (crushed by coal car at the Parkdale

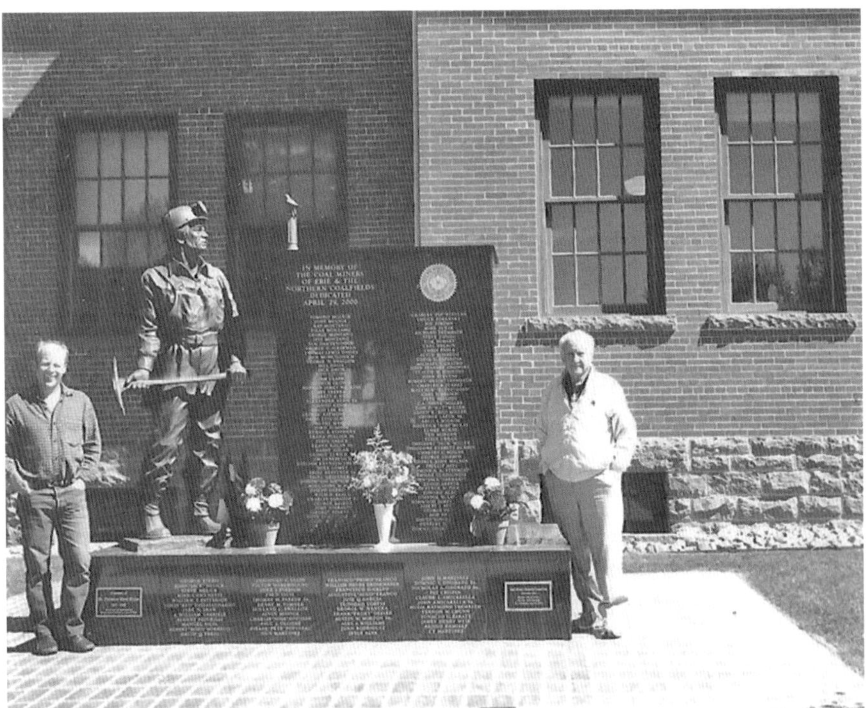


Henry Amicarella and his son, John, stand next to the Coal Miners' Memorial in 2001. The memorial bears the names of four of Henry's siblings: John Sr., Claude, Lawrence and Tony. *Marianna Amicarella collection.*

Mine), Jim Sharratt (cave-in at the Washington Mine), Joe C. Jaramillo (cave-in at the Monarch Mine), Earl Starkey (crushed at the Puritan Mine), T.J. Turpin (crushed at the Puritan Mine) and William Walters (crushed at the Centennial Mine). Some of these names are not on the memorial. Joseph Jordan Jr. died of a heart attack in the portal of Washington Mine as he was getting ready to go home from work. The family remembers the irony of that happening, considering all of the dangers he faced daily on the job. John, Thomas and William Webber of Erie were killed at the Schofield Mine in Utah in 1900.

The following names are listed on the Coal Miners' Memorial in front of Erie Town Hall on Holbrook Street. They are alphabetized here, but they are listed in various blocks on the monument.

John "Jack" Abernathy
Benhamin Abeyta
Edward "Eddie" Abeyta
Joseph G. Abeyta
Lloyd Abeyta
Jesus "Jess" Acosta
Edward Alaux
Johnny Albino
Jesus Alva
Albert Steck Ambrose
Claude L. Amicarella
John Amicarella Sr.
Lawrence Amicarella
Tony Amicarella
Ernst "Swede" Anderson
Ramon Arrendondo
Joe Arroyo
Phillip Asti
James Paul Avila
Ben Ballinger
Frank A. Banyai
Pedro "Pete" Barajas
Rufus M. Barajas
George Baskos
Joe Bell
Pete Bellodi
Jack N. Beltz
William L. Bernal
James A. Berry
Thomas Bertram Jr.
William "Bill" Betz
Aron Billington
Charles Billington
Charles M. Billington
Evertte Billington
Harry Billington
Nick Bokarich
Albert "Bozo" Borrego
Alvin Borrego
Eddie Borrego
Felix Borrego
A. "Tealie" Bottinelli
Allen G. Box
Holland Bracegirdle
Ralph Bracegirdle
Chuck Brack
Gus Brack
Walter Brack
Wilbur Brack
William Brack
James Brennan Jr.
James Brennan Sr.
Joe Brennan
Michael J. Brennan
William R. Brennan
Carl P. Brown

John Brown
Kenneth E. Brown
Richard Brown
William H. "Bill" Brown
Dominic Buffo
John Bujaci
Arturo Calvo
Francisco G. Calvo
Ross P. Calvo
Archie R. Canaday
Fee Columbus Canaday
James Oscar Canaday
Joe A. Cardenas
Joe C. Cardenas
Lee E. Cardenas
William Carrick
Michael Y. Castro
James Charlesworth
Joe Chepelenko
Vernon W. Chunn
William Chunn
Mike Ciocchetti
Nello E. Ciocchetti
Sam Clark
Nick Coloff Sr.
Ysidro T. Comacho
Alfred Cordova
Trinidad Cortez
Pat Crispin
Charlie A. Cruz
John Dale
Thomas Lewis Davies
Bill Davis
David "Singer" Davis
Jack Davis
Jerry Davis*
Sam Dayanovich
Mike DeBaca
Andy Deborski
Leo F. Deborski
Thomas Dominico DeCino
Frank "Froxy" Defler
Max R. DeLiere
Paul L. DeLiere
Jim Denezevich
Gennaro A. DeSantis
Louis "Red" DiGiallonardo
Sam "Shorty" DiLullo
Henry DiMaggio
John Robert Dodds
Robert H. Dodds
Wilfred E. Dodds
Frank Dowling
William Henry Drinkwater
Mike Drovez
Armand Dufor
Theodore Everett Durbin
Alexander Dyet
John Eastenes*
Peter Eichhorn
Clem Epley
Emile Eskro
George Eskro
Eddie U. Espinosa
Ben Espinoza
Samuel Espinoza
Lorenz "Lew" Evans
William M. Evans
Luke J. Evezich
George Farris
J. Candi Fernandez Sr.
George A. Fiechtl
August Figurilli
John Fink
Joe "Cotton" Fletcher
James S. Flowers
Fred Foreman
Benjamin Fragale
Augustine "Augo" Franco
Franisco "Primo" Franco
A.William Gabriele
Joseph Gabriella
Marion Gallegos
Ray Garbiso

Tony Tircio Garbiso
Ben J. Garcia
Daniel Garcia
Jesus G. Garcia
Joseph Q. Garcia
Richard L. Garner
"Shorty" Genoff
John Gentile
John German
Tom Gillies
Arthur Glover
Matthew "Tuffy" Gold
Mike Gomez
Alex N. Gonzales
Bernardo "Ben" Gonzales
Fred Gonzales
Steve Gorce
Dudley F. Gordon
James W. Graham Jr.
John Graham
John A. Grbac
John H. Grbac
Jack Grimson
Tom Grimson
Emilio Guadagnoli
Jim Harris
Miles Harris
Tom Harris
John Hattel
John Hattel Jr.
Fritz Hefftner Jr.
Fritz Hefftner Sr.
Charles E. Hobbs
Leonard W. Hobbs
Robert E. Hobbs
T. Milton Hobbs
John Holley
Tommy "Shorty" Holton
James Henderson Hood Sr.
James H. Hood Jr.
Walter John Hood
Robert Hulsey
James R. Husband
Richard S. Husband
John P. Hutchison
Joseph William Irwin
Charlie Jabot
Thoe Jabot
Joe G. Jacques
Rene Jacques*
Joe C. Jaramillo
Harry Jarvis
Jonas Johnson
David Lincoln Jones
George Raymond Jones
Joe Jordan
Charles R. Juarez
Felix Katich
John Katich
Alexander Kehm Sr.
David Kehm
David Lee Kehm
William "Red" Kellett
Earl Ketsenburg
John Kionka
Eli Kirby
Carroll S. Knoop
Charles Knowles
John Stanley Knowski
Edwin Kolar
Frank Kolar Jr.
Frank Kolar Sr.
Harry Kolar
Fred Korach
Joe Koss
Frank Kostenbater
Frank Kovich*
Alex Kraschuk
Nicholas E. Kreekon
Joe Krpan
John Krpan
Carl Kusick
Lonnie LaFollette
Cornelio M. Lasino

Joe Lasnik
Jack Lasovich
Johnnie W. Lawley
Norwood James Lawley
William "Tup" Lawley
Jesse Earl Layel
George Lazuk Sr.
Adam Leinweber
Phillip Leinweber
John B. Lewis
George Ligeros
Charlie Loasso
P. "Packy" Lombardi
Sabino "Sam" Lontine
Benjamin Lopez
John Lucero
Robert "Bob" Lucock
Paul Lujan Sr.
George Lukich Sr.
George W. Lybarger
Joseph Martain Lyons
George John Machin
Tony Madrigal
Benny Maes
Paul L. Magana
Edward Major
John Major
Joseph Major
Henry Makin
Jack Mamcilovich
Jordy Mann
Bernard R. Manzanares
Dan Marcantonio
Sam Marcantonio
Sam Marino
George Marnich
Cy Martinez
Ed Martinez
Gus Martinez
John D. Martinez
Julian Martinez
Richard Ben Martinez
John Mason
Joe Massaro
Martin H. Mata
George Mathews Sr.
Howard J. Mathews
Frank Mautz
John A. Maxwell
Fred Mazzacco
George Mazzini
George M. "Jiggs" McAfee
John "Jack" McAfee
John Wayne McAfee
Max E. McAfee
Walter G. McAfee
Walter "Shorty" McAfee
R.F. "Pat" McCarthy
Elmer McKay
Roderick "Rod" McKay
Bob McKenzie
Harold Meikle
James Meikle Sr.
John "Jack" Melick
Steve Melick
Pedro "Pete" Mendez
Woodrow Metheny
Wilbur Metz
Frank Mihelcich
Pete Milanovich Jr.
Andrew Miller Jr.
Andrew Miller Sr.
C.A. "Shine" Miller
Charles C. Miller
Chuck Miller
Don D. "Kat" Miller
Elvie Miller
Gillard Miller
Joseph M. Miller
Lloyd R. "Big" Miller
Matt Miller
Thelbert "Jack" Miller
Ray Milo
Mike Milus

George Mircos
Andrew Monares
Joe M. Monares
Joseph F. Monares Sr.
John Monks
George Montano
Jess Montano
Pilar Montano
Ray Montano
Albion Allen Monteith
Nemeczo "Monte" Montemayor
Tony Montour Sr.
Albert Moon
Archie P. Moore
Robert "Bob" E. Moore
Ted Morgan
William Alexander Morrison
William A. Morrison
Austin W. Morton Sr.
John Morton
Alvin Mosnik
Edmond Mosnik
Tony Mosnik
Augustine M. Moya
Ben O. Moya
Tom Mullens
Senon Muniz
Santon Munoz
Pete Murkovich
John Musick
Earl J. Nace
William Nace Sr.
Carl Nelson
John Nenadich
Matthew Nicholas
Ed Oakley
Harry O'Conners
Florencio Ojeda
Ralph Olivas
Domnic V. Onorato Sr.
Nicholas A. Onorato Sr.
Modesto Ornello
James E. Ortega
Paul A. Ortiz
George Otava
Lowe Otava
Hilario Pacheco
Chris Padia
Joe Adam Padia
William Albert Palamet
Anton Panayotoff
Thomas Parker Sr.
Aaron Parkin
Richard Parkin
Peter Pascoff
William T. Patterson
"Til" Pease
Andrew Peich
Nick Perich
John Frank Petras
John Petras Sr.
Julius John Petras
William "Bill" Petras
Claudio Nunzio Petrocco
James Petroff Sr.
Steve Petroff
Lou Phillips
Lawrence W. Pickett
A.M. "Bert" Pierson
Manuel Pina
Albert "Stead" Plakorus
George J. Plakorus
Franklin Platzek Jr.
Frank Platzek Sr.
Fred Platzek
Pierre "Pete" Popineau
LaVoyce Popino
Fred Potts
Pat Powers
Frank C. Protsman
Adelido Quintana
Eli Radjenovich
Rejino Ramirez
Romaldo Razo

Albert Regnier
James Dussart Regnier
Charlie Reise
Angelo Reynoso
George L. Richards
Al Richardson
Henry "Hawk" Rivera
Marcos Rivera
Bill Roberts
Felix P. Robles
Frank "Rod" Rodriguez
Joe A. Rodriguez Sr.
Juan Rodriguez
Anesito Rojas
Richard R. Romero
Peter P. Ross
Julius F. Rossman
Tony Ruiz
Edward "Ed" Russell
Charles Sabados Sr.
Bob Sanchez
Felix Sanchez
Hilberto Sanchez
Nick L. Sanchez Sr.
Peter Sasick
Steve Sasick
Sammy Seidle
Joseph Serola
Bob Shaw
Jabe N. Shaw
Joseph "Joe" Sheader
James Sherratt
Bill Silva
Emilio Silva
Ernst Silva
Manuel Silva
Chuck Simansky
Joseph "Big Joe" Simansky
Claude M. Simmons
John Simpson
Samuel Sims
George G. Sirokman
Anton "Tony" Skriner
Tom Snyder
Tom Somsky
Charlie "Sodi" Sotosek
Frank "Mutt" Sotosek
Nick Spanudakhis*
George Sparakis
John Sparakis
Arthur Spendlow Sr.
Edmund C. Steckler
D.W. "Bill" Stephens
Ethan B. Stephens
Dominick A. Stermlin
Alfred "Jack" Stevens
Charles "Pip" Stevens
John Stevens
Edwin F. Stolns
Herman David Strehlow
"Chub" Summerfield
"Slip" Summerfield
Henry M. Sumner
John Sumner
Mike Sutak
George W. Swallow
Holland C. Swallow
Victor Tafoya
Charles Tangy
Paul Tarasuk
Jack Tierney
E.S. Tilley
Herbert H. Tittle
Jim Todoroff
David Toledo Sr.
Joseph Tomich
Alex Torres
Ed Torres Jr.
Ed Torres Sr.
John Torres
David Q. Trejo
Hugh Raymond Trembath
Hugh Richard Trembath
Richard Trembath

Robert "Brick" Trembath
William Kenneth Trembath
John Tursick
Mike Uitich
Felix Urban
Joseph Urbancich
Jandrya Vagrejvich
Matias V. Valdivia
Frank Valechonchic
Emeterio Valencia
Theodore E. Valin
Pete Velcoff
Mike Vidovich*
Ted Vinovich
Walt Vinovich
Fred Vigil
Richard D. Vigil
Tony O. Villa
Isaac Walters
George W. Waneka
Ramond "Tony" Waneka
Herman James Watkins
James T. Watson
Aim Warembourg
Arthur Warembourg
August Warembourg
Victor Warembourg
James Henry Weir
Albert Wendzel
Joseph F. Wendzel
Everett E. West
Frank Westwood
Thomas Henry Wiggett
Thomas William Wiggett
Alva A. Williams
Arlo A. Wilson
Arthur "Pike" Wilson
James O. Wise Sr.
Rodney R. Woodson
George M. Woolley Sr.
John "Jack" Woolley
Jack T. Wright
Kenneth Wright
Frank Wrobel
Rich Young
Joseph "Joe" Zado
Joseph W. Zaragoza
Ygnacio G. Zarate
Steve Ziggic
Julius "Bud" Zinge
Tony Zlaten

11

# HISTORIC STREET NAMES

Only streets that have historical significance are listed in this chapter. Some streets on the original Erie plat do not exist today: Strong, Mygatt, Fisher and Palmer. Alan Wise of the Wise Homestead Museum believes the streets were never constructed. Hallie Sawyer of Erie Town Hall

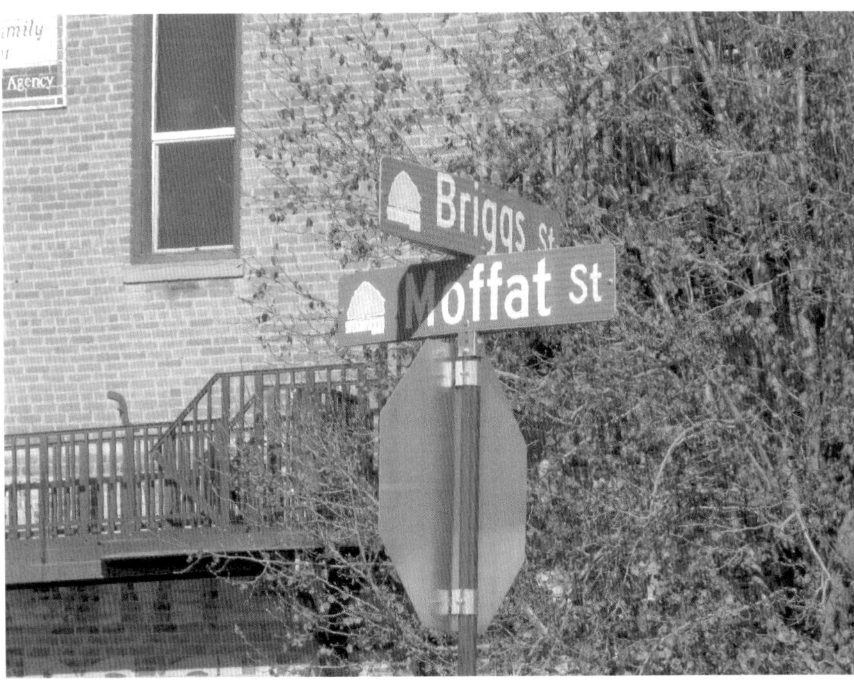

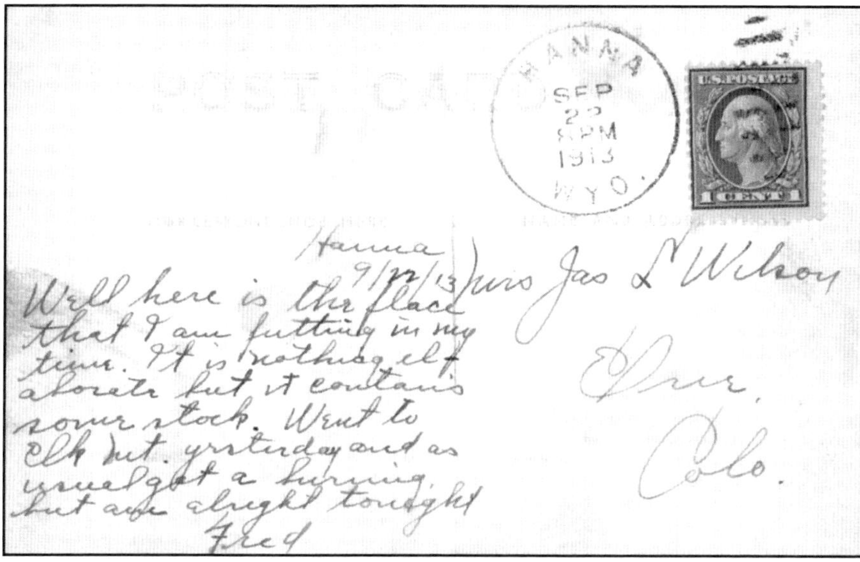


This postcard was mailed in 1913 from Hanna, Wyoming, to Erie by Fred Wilson. Detailed addresses were not necessary. *Sue Larson collection.*

*Opposite*: Street signs began to appear in the Erie area in the late 1940s and early 1950s. The town was so small that house numbers and signs were not necessary as everyone had a post office box. *Penny Stull collection.*

reports an old plat map on file at Weld County that lists East County Line Road as Water Street. Jasper Road used to be Wise Road.

Many old town streets were named after early leaders who influenced the development of the town. Officials of the Boulder Valley Coal Company, attorneys and railroad officials did the original platting of the town, and some streets are named for them.

Streets in Erie Commons are all named after miners. The Miners' Memorial in front of town hall lists many of the names.

Because people received their mail at post office boxes in Thomas Richards's store, they didn't use street addresses. Street signs did not appear in Erie until the 1940s or '50s.

In some cases, more than one name is listed, in others, best guesses are provided.

| Street Names | Probably Named After |
|---|---|
| Agave Way | William Angove, owner of Angove and Hershey feed and livery and grocery store in 1878 (spelling changed) |
| Akers Court | Elmer Fred Akers, doctor |
| Allen Avenue/Court | George Allen, stage stop owner |
| Alva Court | Jesus Alva, miner |
| Ambrose Street | Albert Steck Ambrose, miner |
| Anderson Street/Court | Earl, John and Eric Anderson, three generations of members of Idaho Creek School Board, which originated in 1864 |
| Arapahoe Road | Arapaho Tribe |
| Austin Avenue | Ira Austin, discovered first coal vein in Erie |
| Bain Court/Drive | George Bain, miner and merchant |
| Baker Lane | Jim Baker, Indian scout, mountain man, explorer and Erie's first miner |

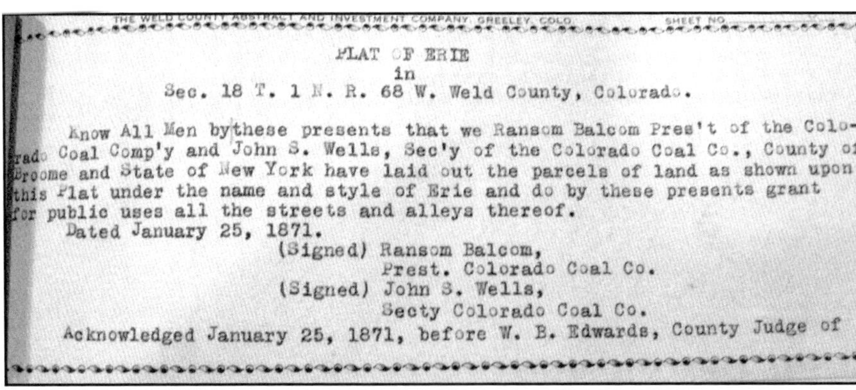
THE WELD COUNTY ABSTRACT AND INVESTMENT COMPANY, GREELEY, COLO. SHEET NO.

PLAT OF ERIE
in
Sec. 18 T. 1 N. R. 68 W. Weld County, Colorado.

Know All Men by these presents that we Ransom Balcom Pres't of the Colorado Coal Comp'y and John S. Wells, Sec'y of the Colorado Coal Co., County of Broome and State of New York have laid out the parcels of land as shown upon this Plat under the name and style of Erie and do by these presents grant for public uses all the streets and alleys thereof.

Dated January 25, 1871.

(Signed) Ransom Balcom,
Prest. Colorado Coal Co.
(Signed) John S. Wells,
Secty Colorado Coal Co.

Acknowledged January 25, 1871, before W. B. Edwards, County Judge of

Ransom Balcolm was just one of many nonresidents of Erie after whom streets were named because of their financial and legal roles in incorporating the town. *Erie Historical Society.*

| Street Names | Probably Named After |
|---|---|
| Balcolm Court/Street | Ransom Balcolm, first president of the Boulder Valley Coal Company, a founder of the Colorado Coal Company and helped plat the Town of Erie. He came to Colorado from New York. |
| Banner Circle | Banner Mine; Banner brothers, owners of a transfer service |
| Baxter Farms Lane | Baxter family |
| Bean Court | August Bean, saloon operator in 1886 |
| Beasley Drive | Beasley family, early homesteaders |
| Bell Drive | Frances Bell, town clerk; Adam Bell, miner; Charles D. Bell, started *Erie-Canfield Independent* |

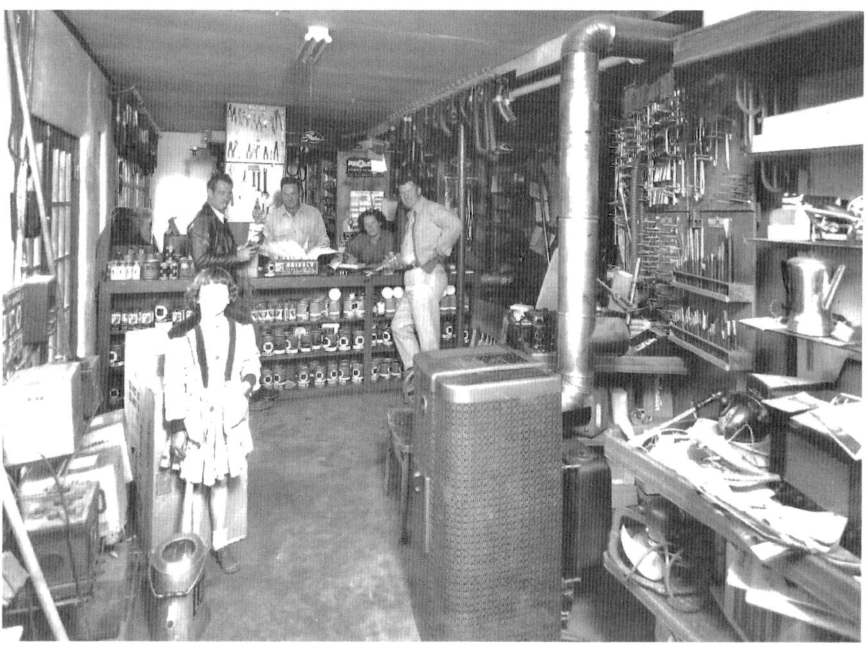

Tommy and Alma Beshear (center) pose for a picture in their store on Cheesman and Briggs Streets with Charles Waneka (left), Claude "Boots" Starkey (right) and Mary Jane Beshear (foreground), circa 1951. *Colleen Waneka Dame collection.*

| Street Names | Probably Named After |
|---|---|
| Beshear Court North/South | Thomas Beshear, owner and operator of a service station and an auto supply and home appliance store |
| Betts Circle | Henry Betts, miner killed at Eagle Mine |
| Billington Drive | Charles Billington, greenhouse operator |
| Bixler Boulevard | Clarence W. Bixler, doctor |
| Block Court | Block, store owner |
| Bonnell Avenue | J. Bonnell, butcher shop owner |
| Brennan Court/Street | Michael J. Brennan, twenty-fourth mayor |
| Briggs Street/South Briggs Street | David Briggs, a New York judge and one who helped establish the town; Henry Briggs, founder of Briggs Mine and official at Boulder Valley Coal Company and Colorado Coal Company. Another possibility is that it was named for Cordelia Briggs van Valkenburg. |
| Brimble Court/Drive | David Brimble, boardinghouse operator in 1886 |
| Carbon Court/Street | Lignite coal/carbon |
| Cardens Court/Place | Cardens family |
| Carlson Avenue | George Alfred Carlson, Colorado governor; Edwin Carlson, miner |
| Carr Street | Robert E. Carr, co-organizer of St. Louis and Denver Land and Mining Company |
| Castro Place | Michael Y. Castro, miner |
| Cheesman Street | Walter Scott Cheesman, businessman and official at Denver Pacific Railroad and one of Erie's founding trustees |
| Clark Court/Drive | Sam Clark, miner killed at Shamrock Mine |

| STREET NAMES | PROBABLY NAMED AFTER |
|---|---|
| Clayton Drive/Way | Clayton Mine |
| Colliers Boulevard | Colliers are coal miners. |
| Columbine Court/Way | Columbine Mine |
| Conrad Drive | Conrad, operator of a service station |
| Conway Court/Street | Conway, owner of jewelry store |
| Cooke Court | W.W. (Myron) Cooke, doctor |
| Cordova Court | Alfred Cordova, miner |
| Cummings Drive | M.S. Cummings, operator of a shoe and boot store |
| Daily Court/Drive | Daily family, early homesteaders |
| Damskee Circle | Theodore Damskee, carpenter in 1898 |
| Davenport Court North/South | Davenport, operator of the City Hotel |
| Davis Court | F.H. Davis, established *Erie Review* |
| Decino Place | Thomas Dominico DeCino, miner |
| Dickens Street | Bill Dickens, stage stop owner |
| Donnelly Avenue/Place | E. Donnelly, operator of a butcher shop in 1878 |
| Dowling Way | Frank Dowling, miner |
| Drinkwater Court | William H. Drinkwater, Erie Trustee and mine owner of the Garfield Mine |
| Eaton Street | Benjamin Harrison Eaton, Colorado's tenth governor (1885–1887) and an irrigation genius |
| Egrew Court | C.P. Egrew, proprietor of Powell House hotel |
| Eichhorn Drive | Peter Eichhorn, miner |

| Street Names | Probably Named After |
| --- | --- |
| Espinosa Street | Eddie U., Ben and Samuel Espinosa, miners |
| Evans Street | Erastus Evans, co-founder of the Colorado Coal Company; John Evans, second governor of Colorado Territory (1862–1865) and financier and developer of the Denver-Pacific Railway and Telegraph Company; William Gray Evans, one of the original incorporators of the Denver and Salt Lake Railway |
| Ferris Drive | George Farris, miner (spelling changed) |
| Fisher Street | on 1871 plat; no evidence of street today |
| Flannagan Court | John Flanagan, mayor (spelling changed) |
| Fletcher Drive | Imo Joe "Cotton" Fletcher, miner |
| Flowers Court | James Flowers, engineer |
| Freeman Court | J.A. Freeman, mayor |
| French Court | C.P. French, Erie's first doctor |
| Gallagher Court | Frank Gallagher, owner of a blacksmith shop |
| Gallegos Lane/Circle | Marion Gallegos, miner |
| Garfield Lane/Road | Garfield Mine |
| Garner Court | Richard L. Garner, miner |
| German Court | Edna German, teacher; John German, miner |
| Gorce Court | Paul or Steve Gorce; Paul operated a saloon, where he was killed. |
| Gordon Drive | Dudley F. Gordon, superintendent at Long's Peak mine, engineer at first power plant and miner |

| STREET NAMES | PROBABLY NAMED AFTER |
|---|---|
| Graham Circle/Way | James W. and John Graham, miners; Thomas E. Graham, owner of the Eagle Mine and chief chemist for the Great Western Sugar Company; George Graham |
| Greening Avenue | William Greening, superintendent of Erie's light and water works |
| Grenfell Court | G.E. Grenfell, president of the board of trustees |
| Grimson Place | John Grimson, miner, died in a cave-in at Lehigh Mine |
| Hamron Court | Andrew Hamron, shoe and boot merchant |
| Harris Court | Jim, Miles and Tom Harris, miners; George Harris |
| Hauck Street | Robert Hauck, early homesteader |
| Hays Court | Roy Hays, bought *Erie Herald* from Olin |
| Hendee Court | A.M. Hendee, Presbyterian pastor |
| Hershey Court | B.F. Hershey, mayor and owner of a feed store |
| High Street/South High Street | Erie Heights neighborhood |
| Hoffman Drive | John Hoffman, established *Western Mirror* |
| Holbrook Street | Holbrook, a New York judge who helped establish the town |
| Holden Court | Eugene Holden, doctor |
| Holeman Drive | Holeman, doctor |
| Hughes Drive | E.C. Hughes, Erie's first mayor |
| Hunt Court | C.R. Hunt, mayor |
| Hunter Court | Hunter, store owner |

| Street Names | Probably Named After |
|---|---|
| Huntley Creek Court | maiden name of builder and developer |
| Irwin Place | David Irwin, co-founder of Erie's first school, which opened on September 1, 1861. |
| Jackson Court/Drive/Lane/Place | Jackson Mine |
| Jacques Way | Joe G. and Rene Jacques, miners; Rene was killed in the Columbine Mine massacre; Joe was also a rancher |
| James Way | Mr. and Mrs. Michael James, donors of site of first school in Colorado (in Erie) |
| Jarvis Drive | Harry Jarvis, miner |
| Jason Court | Jason, doctor |
| Jones Court | Hugh Jones, mayor; Thomas, Ray, owners of the *Erie Independent* |
| Jordan Street | Joe Jordan, miner |
| Kanemoto Lane/Way | James Kanemoto, his family farmed north of Erie |
| Katich | Felix and John Katich, miners |
| Kattel Avenue | E.C. "Judge" Kattel, a New York judge, president of Binghamton Oil Refining Company, co-founder of Colorado Coal Company and one of Erie's founding trustees |
| Kempton Court | Kempton family, early homesteaders |
| Kolar Court | Frank Kolar, mayor; Edward, Frank Jr. and Harry Kolar, miners |
| Koss Avenue | Joe Koss, miner |
| Kramer Drive | Kramer family, daughter of Baxter |

| Street Names | Probably Named After |
|---|---|
| Lakota Ranch Road | named so because property owners had a postcard that showed Lakota (Sioux) tribe members in the area |
| Larson Court | Harry Larson, merchant |
| Lasnik Street | Joe Lasnik, miner |
| Lawley Drive | William Lawley, mayor |

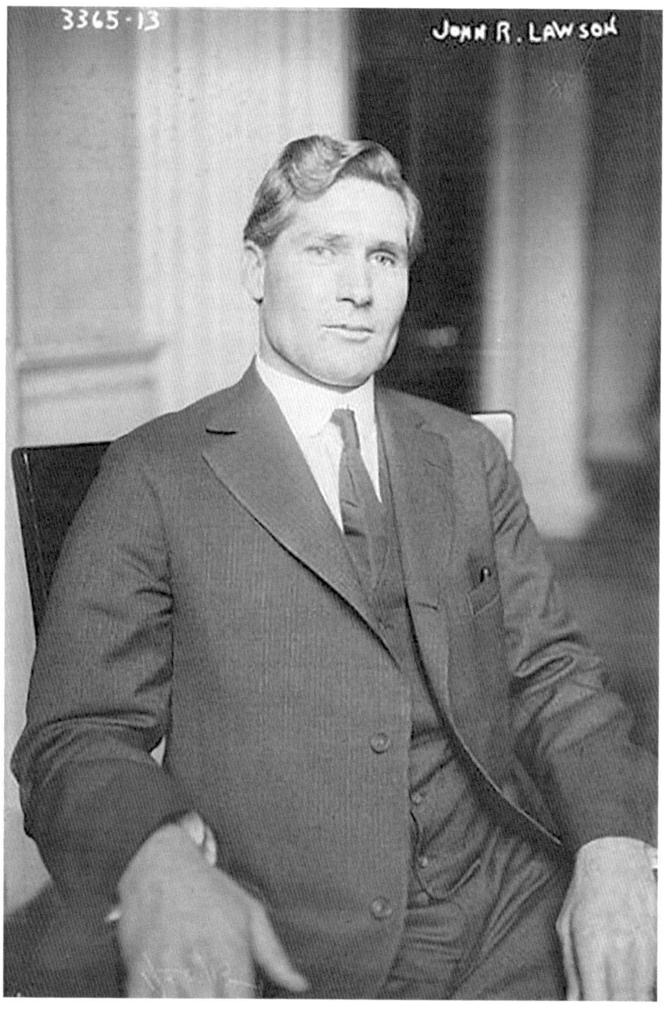


John R. Lawson (1871–1945) was a UMWA leader charged with murder after the Ludlow massacre in 1914. After serving part of his prison sentence, he was found innocent and later became vice president of the RMFC. *Library of Congress.*

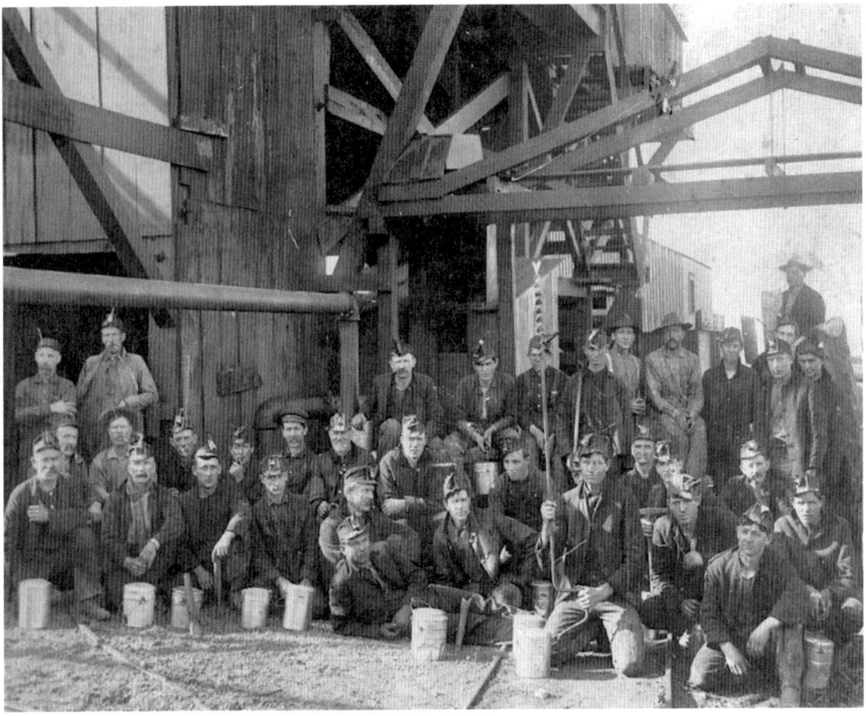

This photo shows a mining crew at the Lehigh Mine between 1902 and 1910. *Lafayette Public Library.*

| Street Names | Probably Named After |
|---|---|
| Lawson Avenue/Court | John Lawson, miner, labor organizer and union leader; Pete Lawson, homesteader |
| Lehigh Circle | Lehigh Mine |
| Leyerly | P.E. Leyerly, owner of the Red & White Grocery in 1936 |
| Leyner Drive | Al Leyner, homesteader; Ella Leyner, milliner |
| Lister Way | Theodore Lyster, assistant cashier at First National Bank of Denver, railroad developer and brother-in-law of Walter Cheesman (spelling changed) |
| Lloyd Way | Lloyd Mine |

| Street Names | Probably Named After |
| --- | --- |
| Lombardi Street | Pasquale "Packy" Lombardi, miner |
| Lucero Place | John Lucero, miner |
| MacCullen Drive | MacCullen, early doctor |
| Madison Drive/Way/Court | M.F. Madison, druggist |
| Main Street/South Main Street | Every town has one. |
| Mann Court | Jordy Mann, miner |
| March Court | Sam March, farmer |
| Marfell Court/Street | Hiram Marfell, homesteader and miner |
| Marshall Court | John Marshall, granted Burlington Northern Railroad right of way on his farm |
| Mason Avenue | Albert Mason, Mayor |
| Mathews Way/Circle | George Mathews, mayor |
| Maxwell Circle/Street | John A. Maxwell, miner |
| Mazzini Street | Reid Mazzini, mayor and miner; George Mazzini, miner wounded at Columbine Mine massacre |
| McAfee Circle/Court | George M. "Jiggs," John "Jack," John Wayne, Max E., Walter G. and Walter "Shorty" McAfee, miners |
| McClure Court/Drive/Way | Thomas McClure, postmaster |
| McDonald Court | John "Jack" McDonald, doctor; George McDonald, barber and justice of the peace; Gertrude McDonald, store and postal clerk; John McDonald, barber and justice of the peace |
| McGregor Circle/South McGregor Circle | McGregor Mine |

| Street Names | Probably Named After |
|---|---|
| Meachum Way | Oscar and Dave Meachum, brothers and owners of blacksmith shop in 1898 |
| Meller Street | George Meller, first president of the board of trustees |
| Miller Street | Miller family, homesteaders |
| Mircos Avenue | George Mircos, miner |
| Mitchell Way | Mitchell Mine |
| Moffat Street | David H. Moffat, railroad official and banker who financed railroad between Denver and Cheyenne; one of Erie's founding trustees |
| Monares Lane | Andrew, Joe M., and Joseph F. Monares Sr., miners and auto shop owners in 1950s |
| Montgomery Drive | C.C. Montgomery, farmer |
| Morris Court/Drive | historical farm |
| Mullens Lane | Tom Mullens, miner |
| Munoz Court | Santos Munoz, miner |
| Mygatt Street | On 1871 plat; no evidence today of street |
| Nelson Street | Carl H. Nelson, miner wounded at Columbine Mine massacre |
| Nonaham Lane | Nonaham, doctor |
| Northrup Drive | Northrup Mine |
| Oakley Lane | Ed Oakley, Erie Trustee and owner of Erie Opera House |
| O'Dell Drive | Jack E. O'Dell, started *Erie Review* |
| Olin Court | Olin, bought *Erie Herald* from McAnally |
| Ortega Court | James E. Ortega, miner |
| Padfield Place | M.W. Padfield, mayor |

| Street Names | Probably Named After |
| --- | --- |
| Palmer Lane | General William J. Palmer, railroad official; Joel Palmer, blacksmith |
| Parkdale Circle/Court/Place | Parkdale Mine |
| Payne Court | Joe Payne, employee of the Works Progress Administration (WPA); Mary Payne, daughter of D.F. Gordon, superintendent of Long's Peak Mine |
| Perry Street | John D. Perry, president, St. Louis, Denver Land & Mining Company; S.M. Perry, president, Northwestern Terminal Railway Co. and one of the original incorporators of the Denver and Salt Lake Railway |
| Petras Street | John Frank, John Sr., Julius John and William "Bill" Petras, miners |
| Phillips Court | Lewis Phillips, miner; Pauline Phillips, pianist at Erie Opera House |
| Pickett Court | Lawrence Pickett, miner |
| Pierce Street | John Pierce, official of the Denver, St. Louis Mining Company and the Denver-Pacific Railway and Telegraph Company; helped establish the town of Erie |
| Plumb Place | Plumb family, early homesteaders |
| Pope Court/Drive | Isaac Pope, saloon owner |
| Potts Court | Fred Potts, miner |
| Powell Street | J.R. Powell, mayor and co-organizer of Long's Peak Coal Company |
| Powers Street | Pat Powers, miner, President of Louisville union local |
| Prince Circle | Prince family, early homesteaders |

| Street Names | Probably Named After |
|---|---|
| Quintana Lane | Adelido Quintana, miner; John B. Quintana, soldier, WWII. |
| Ramirez Circle | Rejino Ramirez, miner |
| Ray Street | On 1889 map, west of Carbon Street, no evidence of street today |
| Reese Court | Thomas Reese, supervisor, Northwestern Mine |
| Reliance Court/Drive/Place | Reliance Mine |
| Richards Court | Tom Richards, miner and merchant |
| Risse Court | John Risse, street commissioner and wagon maker |
| Rivera Court | Henry "Hawk" Rivera, Marcos Rivera, miners |
| Romero Lane | Dora Romero, pioneer; Richard R. Romero, miner |
| Rossman Lane | Julius F. Rossman, miner |
| Sanders Circle | W. Sanders, county clerk; Edward Sanders, blacksmith and manufacturer of coal cars and cages |
| Seidler Court | A. Seidler, operated clothing store |
| Serene Court/Drive | Serene mining camp at Columbine Mine |
| Shuttleworth Drive | John Shuttleworth, mayor |
| Simmons Court/Street | E.W. Simmons, Episcopal pastor |
| Smith Circle | E.A. Smith, co-founder of the Colorado Coal Company (most likely named for him); Walter W., doctor; Smith family, homesteaders; Smith, owner of hardware store |

| Street Names | Probably Named After |
|---|---|
| Southard Street | Samuel H. Southard, town clerk, justice of the peace, livery teams operator |
| Stanley Court/Drive | Clyde Stanley, established *Erie News* |
| Starkey Court | Lewis Starkey, miner |
| Stevens Circle/North/South | Alfred Stevens, mayor |
| St. John Street | Twombly St. John, merchant |
| Stockton Drive | Guy C. Stockton, school principal |
| Strong Street | On 1871 plat; no evidence of street today |
| Summerfield Court | Robert, railroad section foreman; "Chub" and "Slip" Summerfield, miners |
| Tanaka Drive | Tanaka family, farmers |
| Taylor Lane | Job O'Taylor, miner, died in Parkdale Mine |
| Telleen Avenue | J.B. Telleen, developer |
| Turpin Way | T.J. Turpin, machine operator who died at Puritan Mine |
| Tyler Place | Tyler family, early homesteaders |
| Tynan Drive/Court | Tynon Mine (spelling change) |
| Valencia Circle | Emeterio Valencia, miner |
| Walker Street | Franklin Walker, mayor |
| Walters Drive | Early settlers |
| Washburn Avenue/Court/Street | Sarah Washburn, proprietor of Hotel Powell |
| Webber Court | Webber brothers, miners |
| Wells Street | John S. Wells, secretary of Boulder Valley Coal Co., co-founder of Colorado Coal Company |

| Street Names | Probably Named After |
|---|---|
| Westin Drive/Circle | Dr. H.L. Weston (spelling changed) |
| Westwood Place | Frank Westwood, miner |
| Wharton Court | Joseph J. Wharton, president, board of trustees |
| Whiles Court | William Whiles, mayor |
| White Lane | Eben White, stage stop owner |
| Wiggett Court | Thomas Henry and Thomas William Wiggett, miners |
| Williams Court | John T. Williams, merchant and reverend |
| Wilson Circle | J.L. Wilson, Francis W. Wilson, and Fred R. Wilson, mayors; Arlo A. and Arthur "Pike" Wilson, miners |
| Winslow Court | Winslow, G.A., co-owner of hardware store |
| Wise Road (now Jasper Road) | Oliver Ellsworth Wise, homesteader, blacksmith, settled Canfield |
| Woodson Drive | Bob Woodson, owner of a blacksmith shop |
| Woodward Street | Stillman Woodward, doctor |
| Young Court | Robert W. Young, president of the Erie Board of Trustees; Rich Young, miner |
| Zodo Avenue | Joseph "Joe" Zado, miner (spelling changed from Miners' Memorial) |

12

# MAYORS

When Erie was first established, a board of trustees, headed by a president, provided leadership. In 1881, Erie still had a board of trustees, and it elected a mayor. It continues with that structure today. Mayoral terms have been established as two years, running from April to April. Mayors with asterisks by their names are buried in Erie Mount Pleasant Cemetery.

| Presidents of Boards of Trustees | Tenure |
| --- | --- |
| George Meller | 1874–1875 |
| John T. Williams* | 1875–1876 |
| Richard J. van Valkenburg | 1876–1877 |
| G.E. Grenfell* | April–May 1877 |
| Joseph J. Wharton | 1877–1879 |
| Robert Young | 1880–1881 |
| Richard J. van Valkenburg | April–May 1881 |

| Mayor | Tenure |
|---|---|
| E.C. Hughes | September 1881–1882 |
| Richard V. van Valkenburg | 1882–1883 |
| Hugh Jones* | 1883–1885 |
| G.E. Grenfell* | 1885–1887 |
| J.A. Freeman | 1887–1888 |
| J.L. Wilson | 1888–1891 |
| G.E. Grenfell* | 1891–1892 |
| Hugh Jones* | 1892–1893 |
| J.L. Wilson | 1893–1894 |
| Thomas Morgan | 1894–1896 |
| J.R. Powell | 1896–1898 |
| William Nicholson | 1898–1900 |
| Alfred Stevens* | 1900–1902 |
| Albert Mason* | 1902–1903 |
| George Mathews | 1903–1904 |
| B.F. Hershey | 1904–1906 |
| George Mathews | 1906–1907 |
| William Nicholson | 1907–1910 |
| John Shuttleworth | April–November 1910 |
| William Nicholson | November 1910–1914 |
| George Mathews | 1914–1915 |
| William Nicholson | 1915–1921 |
| J.L. Wilson | 1921–1922 |
| John Flanagan | 1922–1923 |

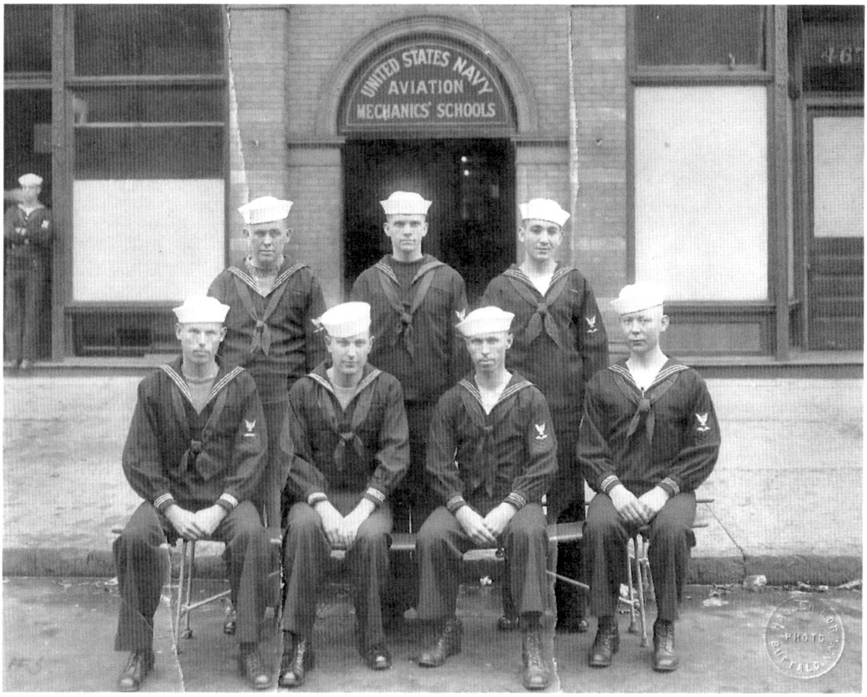


Francis Wilson (back row, left) joined the U.S. Navy and attended a technical school in Buffalo, New York. *Ray Wilson collection.*

| MAYOR | TENURE |
|---|---|
| M.J. Brennan | 1923–1924 |
| William Whiles | 1924–1928 |
| C.R. Hunt | 1928–1929 |
| William Nicholson | 1929–1930 |
| Francis A. Wilson | 1930–1932 |
| Fred R. Wilson | 1932–1934 |
| Ray Jones | 1934–1936 |
| M.W. Padfield | 1936–1940 |
| Charles J. Elzi | 1940–1942 |

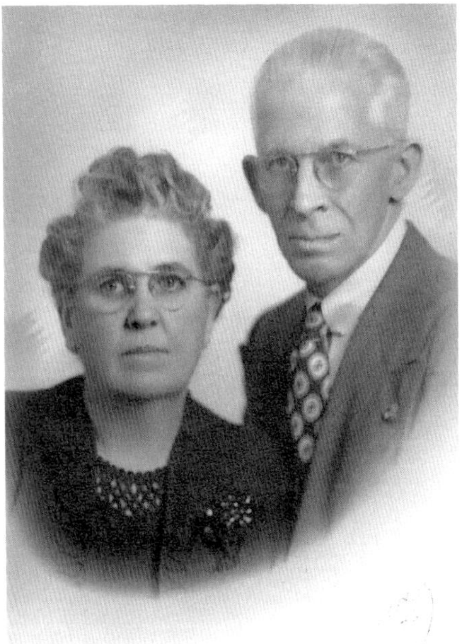

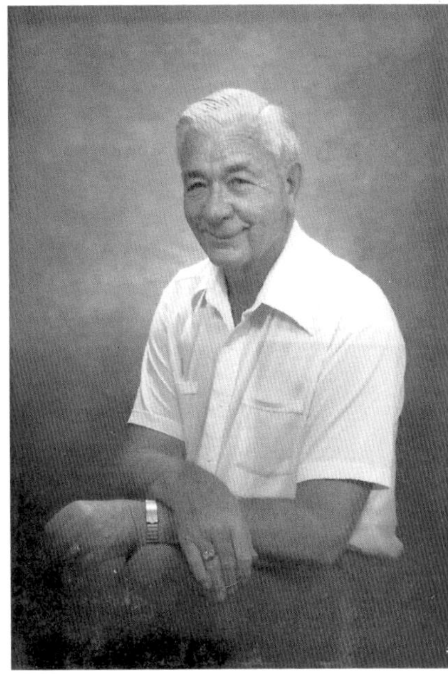

*Left*: Rhoda and Frederick Wilson. Rhoda lived to be 103. *Ray Wilson collection.*

*Right*: Sheldon Allan. *Carolyn Allan McCarthy collection.*

| Mayor | Tenure |
|---|---|
| M.W. Padfield | September 1942–1943 |
| Frank Kolar | 1943–1944 |
| Charles J. Elzi | 1944–1946 |
| Roger Chrisman | 1946–1947 |
| Fred Hasenack | 1947–1948 |
| Fred R. Wilson | 1948–1950 |
| William H. Carter | 1950–1954 |
| Franklin Walker* | 1954–1955 |
| Fred R. Wilson | 1955–1957 |
| Reid A. Mazzini | 1957–1959 |

| Mayor | Tenure |
|---|---|
| Joseph Jordan | 1959–1960 |
| Sheldon Allan | 1960–1966 |
| Harold Conroe | 1966–1968 |
| John H. "Jack" Findley | 1968–1970 |
| Reid Mazzini | 1970–1972 |
| William Kenneth Trembath | 1972–1973 (May–November 1972) |
| William Lawley, pro tem, acting mayor* | 1973–1974 (served remaining 1½ years |
| Harlan E. Brock | 1974–1980 |
| Harold G. Stewart | 1980–1982 |
| David Louis | 1982–1986 |
| Victor F. Smith | 1986–1988 |
| Charles D. Dolphin | 1988–1990 |
| Michael Woodruff | 1990–July 1994 |
| Victor F. Smith | September 1994–2000 |
| Tom Van Lone | 2000–2002 |
| Barbara Connors | 2002–2004 |
| Andrew Moore | 2004–2010 |
| Joe Wilson | 2010–2014 |
| Tina Harris | 2014–present |

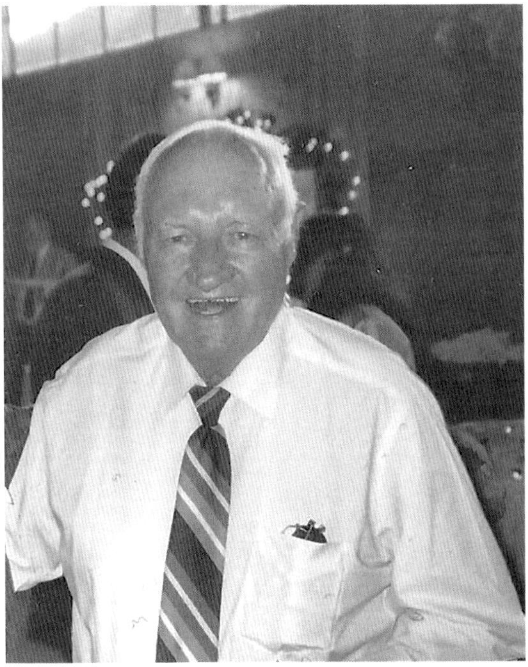

*Left*: Harold Conroe. *Harold Conroe collection.*

*Below*: This 1960s trustees meeting shows *(left to right, standing)* Donald Zaruba, Jack Miller, Sheldon Allan, Jack Stevens, Micky Surtak, Joey Jordan, *(left to right, leaning)* Manuel Pacheco, Charles Stevens and Jim Waneka. *Carolyn Allan McCarthy collection.*

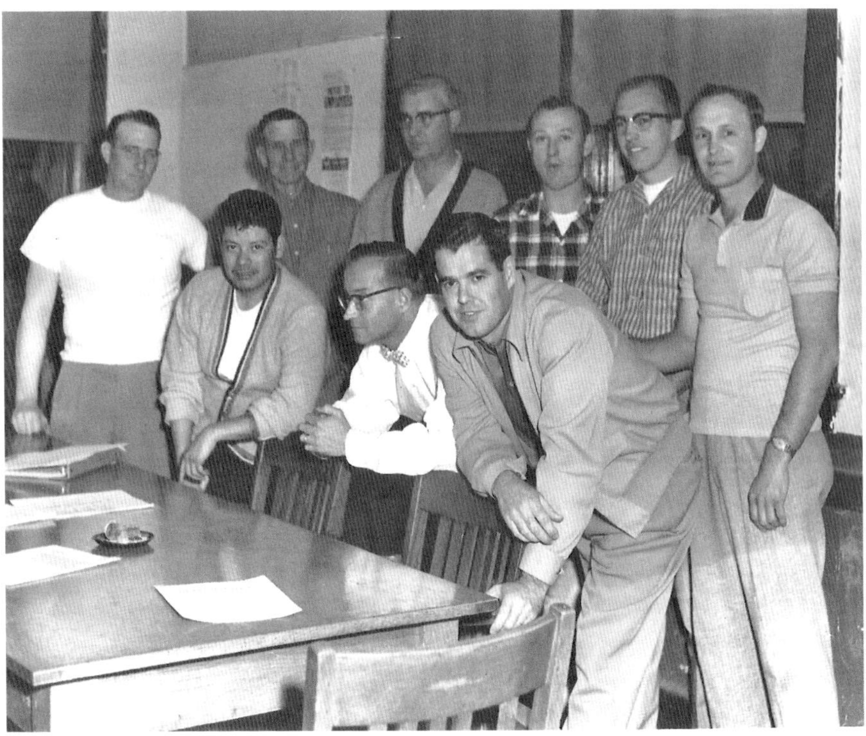

13

# OUT OF THE COAL DUST

From 1919 to 1939, Colorado's population changed very little; between 1920 and 1930, it increased by fewer than 100,000 people. Between 1930 and 1940, it grew by around 88,000 people. During the coal boom between 1860 and 1919, the state's population grew by about 100,000 residents per decade. Erie's population mirrored these trends, with the town not reaching 1,000 residents until the 1940s.

On the west side of Highway 287 at Arapahoe Road (Nine-Mile Corner), two stone pillars mark the gateway to the Road of Remembrance. Following World War I, American Legion members decided to beautify Arapahoe Road to honor military veterans. Arapahoe was a well-traveled route from Denver to Boulder before U.S. 36 was completed in 1952. The plan was to plant one thousand trees along Arapahoe all the way to Boulder, with small parks at each intersection. A captured German field gun and a flagpole were erected on a plot at one intersection. The Boulder Lions Club placed a plaque on the northernmost pillar and buried a metal box with documents, newspapers and the names of club members.

The Great Depression from 1929 to 1939, the Dustbowl of the early 1930s and World War II had significant impacts on Erie's development. Roosevelt's New Deal legislation brought changes for miners and other Colorado workers. The Depression brought attention to the status of labor. Union membership increased. The federal administration was convinced that unfair labor practices contributed to the economic distress of the nation.

After World War II, Erie annexed adjacent lands that showed potential for growth. Even before that, as Erie expanded west, residents expressed their desire to change county boundaries and move all of Erie into Boulder County. Arguments surrounding this issue still exist.

Colorado saw increased production from the uranium boom of the 1940s, the aerospace industry, the Rocky Flats Nuclear Weapons Plant, the Rocky Mountain Arsenal, the Air Force Academy and the increased funding of research for national defense.

In 1956, Erie installed a new sewer system. In the 1960s, natural gas lines were brought in for household use.

Also in the 1960s, Erie annexed Canfield and other Boulder County communities. It also became home to the Mountain View Dragway, a National Hot Rod Association and American Hot Rod Association–sanctioned drag strip. The strip was located across Erie Parkway from Erie High School. Today, you can still see the track as it runs from northeast–southwest toward Weld County Road 5. The two trees across from the football field marked the starting line. The track was noted for its remarkable shutdown area that went on long after the finish line and was graded to help

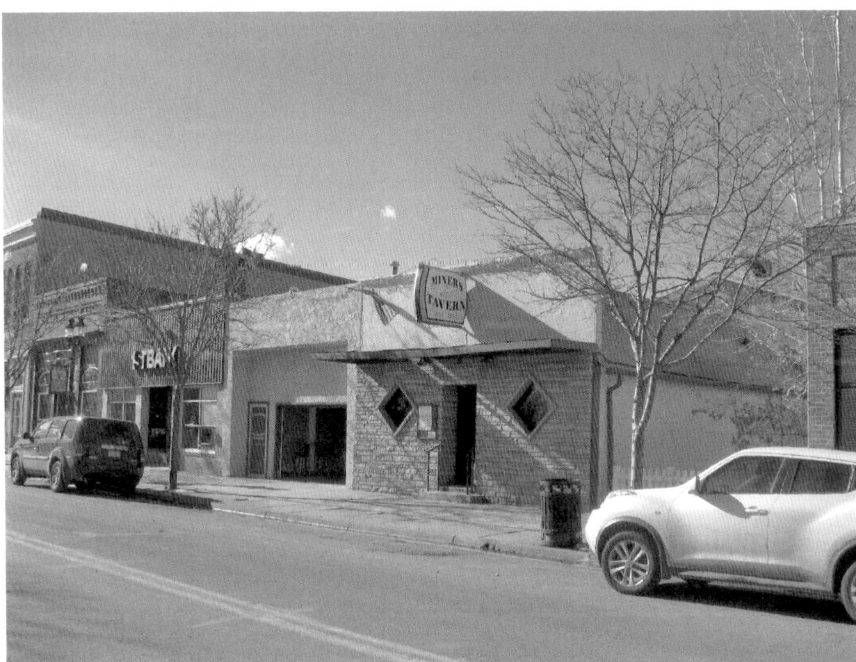

The Miners Tavern on Briggs Street has served Erie since 1926. *Author's collection.*

cars slow down. The strip offered a beautiful view of the Front Range. It was the first racing venue in Colorado to provide covered pits with electrical power outlets. Al Bockla was the first to break the two-hundred-mile-per-hour barrier in Colorado at the Dragway. Albert Broncucie owned and operated the strip, later called Mile High Speedway. In the 1980s and '90s, the track was used as a private airstrip.

In the 1970s, Erie had more bars than churches. Residents reported that it got pretty wild on Saturday nights.

On May 26, 1971, Erie was host to its own mini-Woodstock at Erie's Mountain View Dragway. By then, the promoters referred to the Mountain View Dragway as Erie Raceway. Among the rock stars who played at the concert were the Allman Brothers, Big Brother and the Holding Company, the Youngbloods, Jeffrey Cain, Fresh Air and Friends for Now. Erie residents Debbie Langerak and Patty Bray suggested that Colorado concert promoter Barry Fey may have been involved with this event; however, it was listed as part of the Chicken on a Unicycle promotions made famous by Bill Graham.

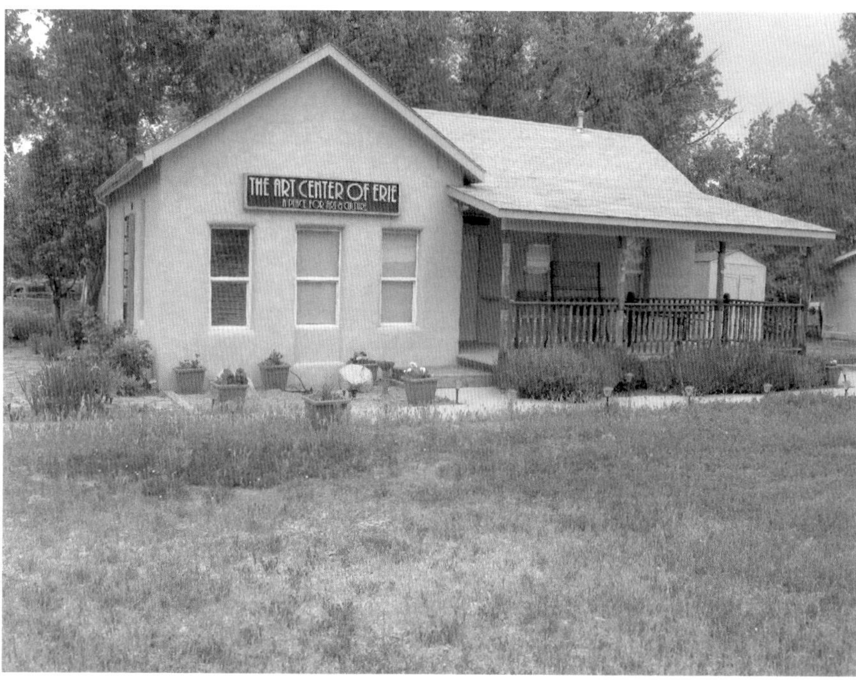


The Lorraine David Children's Library building is now home to the Arts Coalition of Erie. *Author's collection.*

In 1979, Erie Airpark was annexed into Erie, the first of many subdivisions in Erie expansion. Most of the houses suggested affluence, and many residents owned airplanes.

In the 1990s, over ten housing developments were constructed in Erie, bringing in thousands of new residents.

In 1996, the Lorraine David Children's Library was founded at 625 Pierce Street, and it served as Erie's first library; however, it didn't open until 2002. In 1992, Lorraine conceived the idea that if she started a children's library, the town would eventually get a traditional library. The involvement of Annette Francke is found in a few sources.

The 840-square-foot building was moved to Erie from Boulder, where it had existed at 925 Pearl Street since 1874. The last use for the building in Boulder was for Jay's Pasta. Robert Shonkwiler moved it at his own expense as a donation to the Erie Historical Society. Originally, it was a three-room brick cottage built by Charles and Fannie Boettcher. Charles was a wealthy silver magnate. When the building was moved to Erie, it had a cement/stucco covering over the bricks, and that covering still exists today. Lorraine wanted to restore the building to its original brick style.

A committee of about forty people remodeled the building to make it more suitable for a library. They furnished it, added curtains, installed computers and purchased supplies. Lorraine died three days after the library opened in 2002.

In 2003, students from the fifth grade at Erie Elementary School created a peace garden at the library under the direction of teachers Judy Solano and Leigh Carpenter. The project was the outcome of an exercise in which the students discussed their feelings about the September 11, 2001 terrorist attacks on the World Trade Center in New York City. Teachers, students and parents raised money, bought plants, organized a planting day and helped to create pavers that are still displayed proudly on the west side of the building. Various businesses also donated funds to the project, and Erie High School's Educational Foundation Association and National Honor Society contributed to the project's success. Within the garden are bricks bearing "peace" in many different languages, some with the flags of different nations. A peace pole was installed with the words "may peace prevail on earth" in eight languages. Some of the relics and statuary adorn the new peace garden at the Erie Community Library.

Erie library officials and citizens were not happy about how their property taxes were being used by the Weld County Library District, and they fought for years for secession from the district. They believed the funds were being used for other county libraries.

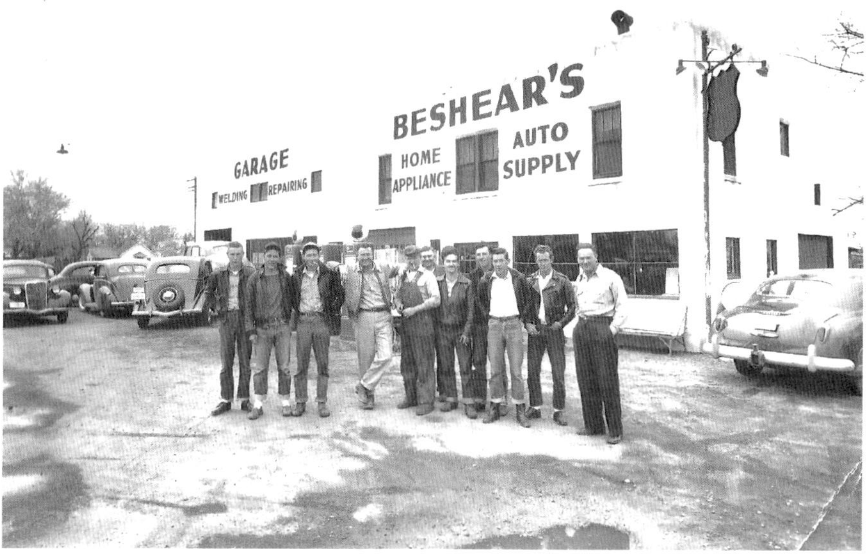


*From left to right*: Billy Jordan, Billy Simansky, Glenn Hobbs, Claude Starkey, Pete Streckler, Alvin Mosnik, Jim Harris, Edmund Mosnik, Clifford Hobbs, Charles Waneka and Tommy Beshear at Tommy's store on Cheesman and Briggs Streets, circa 1951. *Colleen Waneka Dame collection.*

In February 2008, Lorraine David Children's Library president Pat Bundy and library workers worked with volunteers from Pathway to Spirit to load books, computers, furniture and supplies into a truck to be driven to the Pine Ridge Oglala Lakota Reservation in South Dakota. The Pathways program was directed and presided over by Carmeen Klausner. Other books and supplies were moved to the new Erie Community Library.

As recently as 1999, visitors to the town of Erie were surprised to see unpaved, red dirt streets. People said Erie looked like something out of the past—a town forgotten. It was not the Erie of old with wooden boardwalks and hitching posts for horses. The sidewalks were in, and cars were parked along the streets. However, many of the old miners' cabins and commercial buildings still gave the town a look of the Wild West. The project to complete the paving of Erie's streets was completed in 2000. Some report that Leon A. Wurl's ashes were mixed in with the asphalt. He was an influential figure in the later development of Erie.

In 2000, the current Erie Chamber of Commerce building was renovated.

In the fall of 2005, the new Erie High School opened on the northeast corner of Erie Parkway and Weld County Road 5.

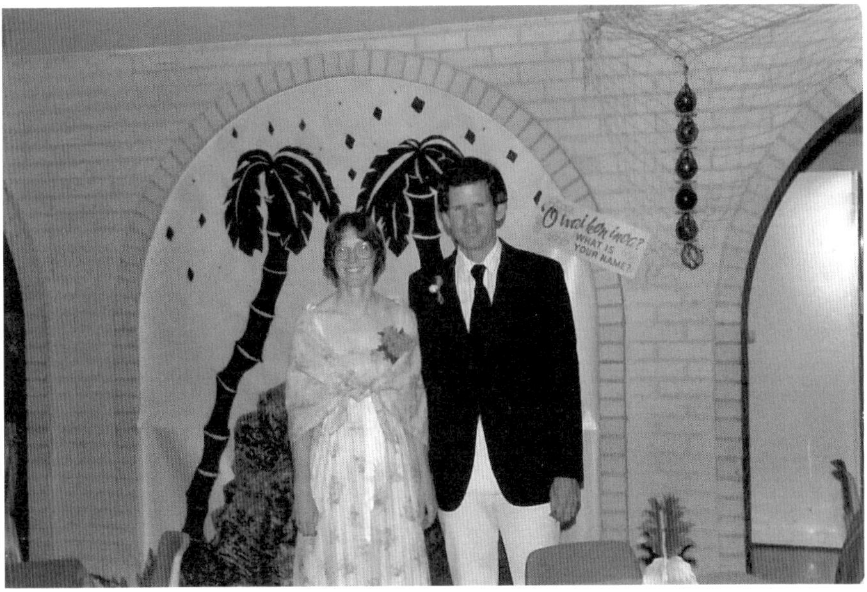

Bill and Susan Pedrick chaperone an Erie High School prom in 1981. *High Plains Library District.*

In July 2007, the Wise Homestead Museum opened to the public. Dr. Sarah Wise, great-granddaughter of the original owner of the property, donated the homestead as a museum to the Erie Historical Society. She lives there today, following a successful teaching and research career. The homestead dates back to 1869, when Oliver E. Wise; his wife, Adeline; and their three children, William O., Joseph O.V. and Ada, came across the plains from Wisconsin to homestead in Colorado. Oliver built a small log cabin and began to farm the land. In 1870, he began building a permanent farmhouse. He operated a farm, a large mill, a grain storage elevator, a blacksmith shop and the Canfield General Store. This property was located in Canfield, which is now part of Erie. Coal was discovered on the property.

On January 12, 2008, the Erie Community Library opened at 400 Powers Street. It is a full service library for all ages. Its collection consists of nearly sixty thousand items and focuses on popular materials in all areas (books, cassettes, CDs and DVDs). The branch maintains a large selection of popular, high-demand fiction titles known as "Express Books" that are available for checkout immediately. In addition, the branch is a wireless hotspot, receives daily courier service and has access to many online databases. It is a branch of the High Plains Library District.

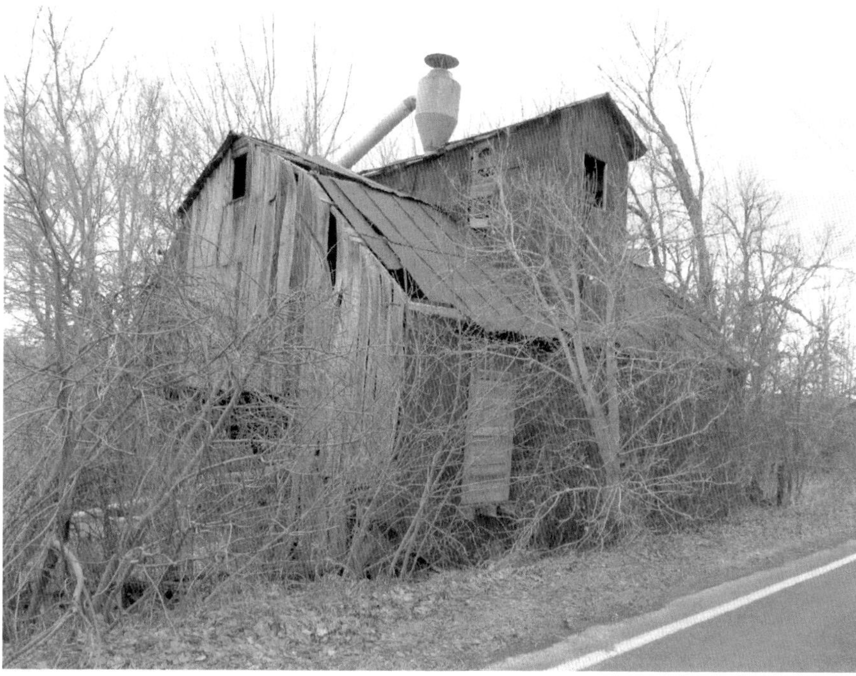

The flour mill on the Wise Homestead still stands today, and it is being considered for relocation. *Erie Historical Society.*

In 2008, the Erie Community Center opened at 450 Powers Street. This sixty-three-thousand-square-foot facility has a lap pool, a leisure pool with play features, a whirlpool, a two-story figure-eight pool slide with a splash tank, a lazy river, an outdoor water feature and a spray area. It houses a gymnasium fitness area on two levels, climbing walls, an indoor running track, racquetball courts, a child care facility with an adjacent playground and an indoor playground and a teen room. It has several meeting rooms and a kitchen as well.

The Colorado National Golf Club opened in 2009 and has earned awards as Colorado's top Front Range golf course. It also serves as the training course for the University of Colorado men's and women's golf teams. This family-oriented recreation spot provides a junior golf program, excellent eateries and venues for corporate events, banquets, weddings and other special occasions. It is located in Vista Ridge on Vista Parkway.

On April 3, 2010, Erie Community Park opened. This approximately forty-one-acre, $6.2-million park is located adjacent to the Erie Community Center on the northeast corner of Erie Parkway and County Line Road.

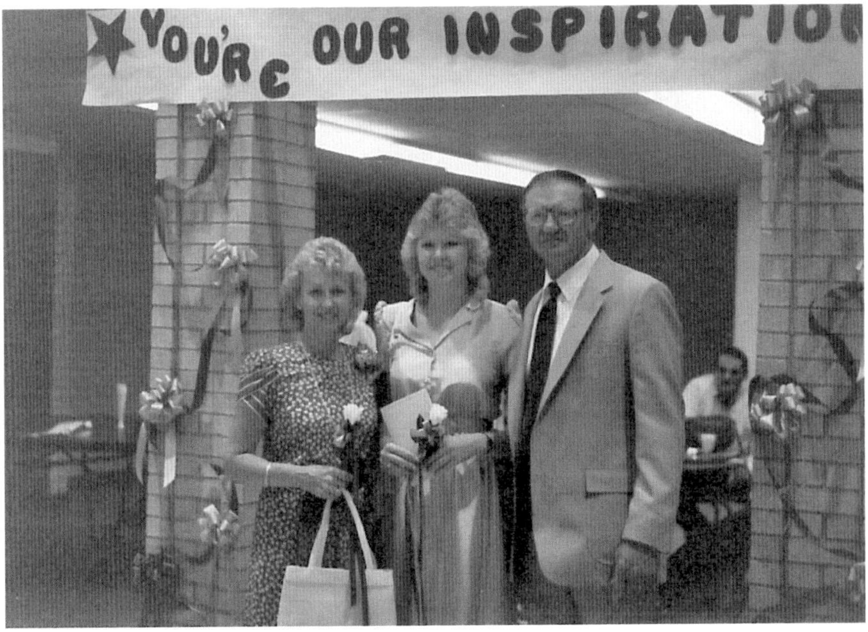


Linda Cleland graduates from Erie High School in 1985 and poses for a photo with her parents, John and Shirley. *High Plains Library District.*

Phase one construction included four lighted baseball/softball fields, a multipurpose field, four lighted tennis courts, a concession stand, three playgrounds and a picnic shelter. The park was funded through the Town of Erie Park Impact Fund, the Tree Impact Fund, a $200,000 grant from Great Outdoors Colorado and a $175,000 grant from EnCana Oil and Gas USA, Inc.

Also in 2010, St. Luke Orthodox Christian Church opened on the southwest corner of Austin Avenue and County Line Road.

Erie Parks and Recreation opened a skate park in the Erie Community Park in 2013. It was designed by SITE Design Group and built by California Skateparks. This Street League Skate Plaza features stair gaps, flat ledges, hubba ledges, grind rails and banks. Involved in negotiating some of the features were Nyjah Huston, Chris Cole, Shane O'Neill, Paul Rodriguez and Tom Asta. Square State Skate is scheduled to provide classes and other events.

The town of Erie has developed miles of safe, convenient walking and bicycle trails. Funded by a half-million-dollar grant from the Great Outdoors Colorado project, the Coal Creek Rock Trail is designed to connect Erie

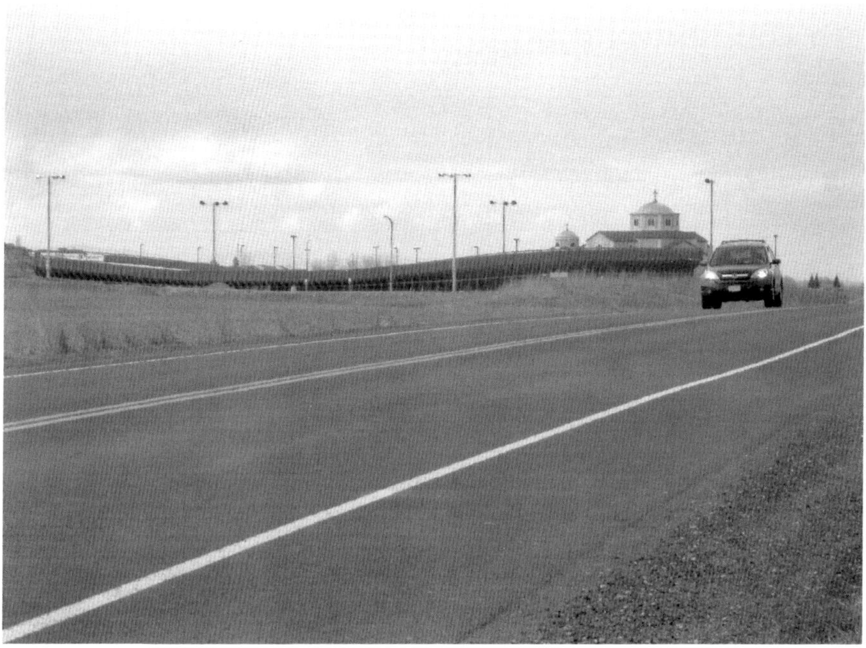

Erie's skyline has changed over the years. Pictured are the Velodrome, which opened in 2015, and St. Luke Orthodox Christian Church, which opened in 2010. *Author's collection.*

with Lafayette, Louisville, Superior, Broomfield and parts of unincorporated Boulder County through a twenty-four-mile loop trail.

The Boulder Valley Velodrome is located on East County Line Road. This 250-meter racing oval is touted as a world-class Olympic-caliber facility available to recreational riders as well as those training for Olympic races.

Erie's landfills provide important revenue for the town. Efforts to expand the landfills are highly controversial. Many historic mine sites have been buried. A number of historic landfills are no longer in use.

Erie's population was 645 in 1877, with five major mines, fourteen businesses, two churches, two hotels and several boardinghouses. Between 1990 and 2008, Erie's population grew over tenfold, to almost 15,000 residents. The following shows how Erie's population grew from 1880 to 2015.

## Erie's Population Growth 1880–2015 (United States Census and Town of Erie figures)

| Year | Population | Year | Population |
|---|---|---|---|
| 1800 | 358 | 1970 | 1,090 |
| 1890 | 662 | 1980 | 1,254 |
| 1900 | 697 | 1990 | 1,258 |
| 1910 | 596 | 2000 | 6,290 |
| 1920 | 697 | 2008 | 14,958 |
| 1930 | 930 | 2010 | 18,135 |
| 1940 | 1,019 | 2011 | 18,900 |
| 1950 | 937 | 2013 | 20,500 |
| 1960 | 875 | 2015 | 22,000+ |

Since the 1990s, Erie has been regarded as a bedroom community for Boulder County and Denver. Commercial development began along Interstate 25. Plans were being made to expand Erie to a population of fifty thousand residents.

In 2010, the median age was 32.5. There were 6,485 households. The median household income was an astonishing $84,284, compared to $47,827 in Denver. In January 2011, the number of households rose slightly to 6,620. The median household income rose to $99,804.

Between 1874 and 2008, Erie annexed 10,894 acres. In 2014, Erie had a planning area of forty-eight square miles—about the size of the city of San Francisco. It has incorporated nineteen square miles, including 346 acres of parks and open space. It has 7,150 households, with another 7,000 planned.

The Colliers Hill, Erie Highlands and Golden Run developments will bring many more residents to Erie. With this growth will come more businesses. Erie plans to open a supermarket on Highway 7 in the near future. There has been talk about bringing the railroad back to Erie—a commuter line between Erie and Interstate 25 that will connect with a main line to Denver.

Two elementary schools opened in 2011. Aspen Ridge is a charter school and is located at Austin and Lloyd Avenues. Red Hawk Elementary

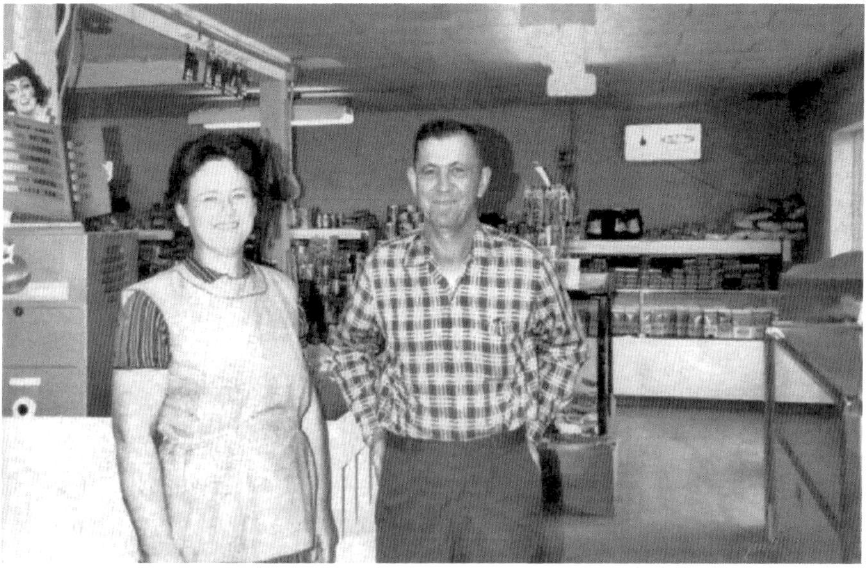

Tony and Josie Hepp owned and operated Anthony's Market from the late 1960s until 1970. They leased the building to other merchants at the northeast corner of Cheesman and East County Line Road, the current site of the Great Western Bank. *James Sites collection.*

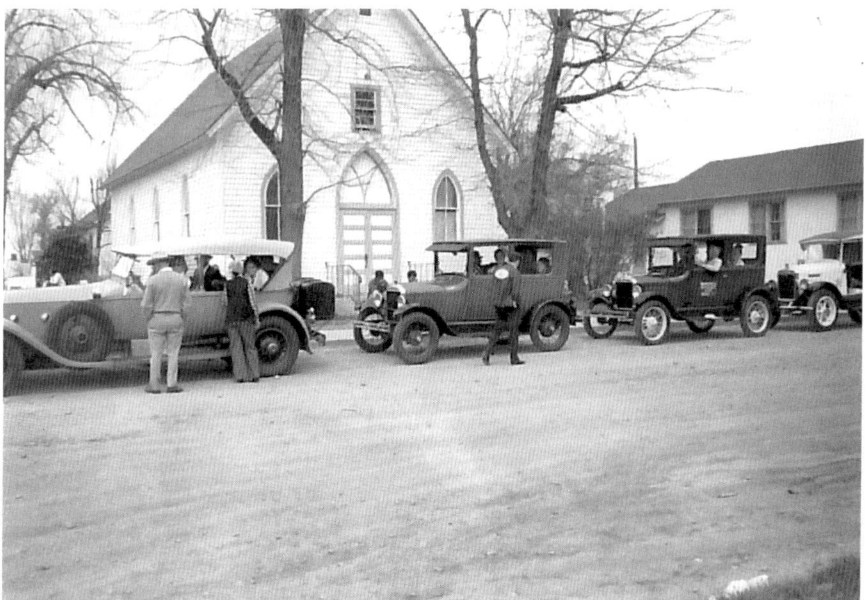

Antique cars line up in front of St. Scholastica Catholic Church in preparation for the 1976 Erie High School homecoming parade. *High Plains Library District.*

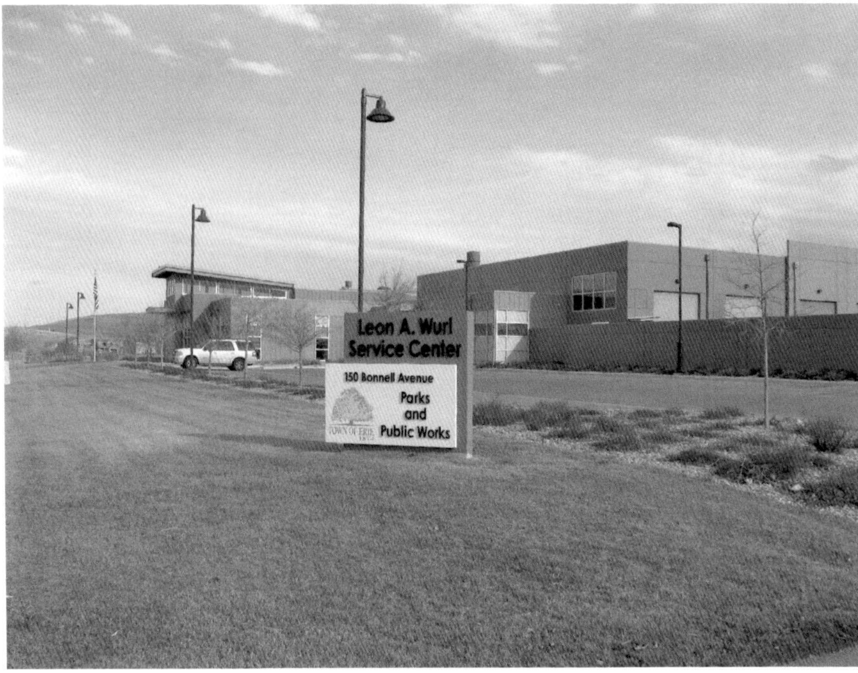


Leon Wurl was the town administrator who brought Erie out of the coal dust and into the twentieth century by transforming the Lincoln School into town hall, building water and sewage plants, paving Old Town streets and encouraging residential and commercial growth. *Author's collection.*

School is west of County Line Road at Telleen and Meller Streets. Another school is being planned for the Flatirons Meadows community. The Calvary Bible Church was built in 2014.

The future of Erie remains to be seen. It will likely continue to be a bedroom community that serves areas where employment opportunities are found. While millions of tons of coal are still mined in Colorado every year and 129 billion tons of potentially minable coal underlies much of the state, the likelihood of the return of mining in the Erie area is slim. Oil and natural gas have virtually replaced coal as an energy source, and their production promises to continue in Weld County.

# BIBLIOGRAPHY

Dyni, Anne. *Erie, Colorado: A Coal Town Revisited.* Erie, CO: Town of Erie, 2001.

Kozma, Jane, Jane Marshall, Mary Montano and Dave Stone. "Saint Scholastica Church, 1899–1999: A Journey of Faith." St. Scholastica Church, Erie, CO, 1999.

May, Lowell, and Richard Myers, eds. "Slaughter in Serene: The Columbine Strike Reader." Bread and Roses Cultural Center, Denver, 2005.

Michener, James A. *Centennial.* New York: Random House, 1974.

Muncy, Robyn. *Relentless Reformer: Josephine Roche and Progressivism in Twentieth-Century America.* Princeton, NJ: Princeton University Press, 2014.

Ogle, George. *The Price of Colorado Coal: A Tale of Ludlow and Columbine.* Bloomington, IN: Xlibris Corporation, 2006.

Perry, Phyllis. *Speaking Ill of the Dead: Jerks in Colorado History*. Guilford, CT: Global Pequot Press, 2011.

Simmons, R. Laurie, and Thomas H. Simmons. *Erie, Colorado Historical Buildings Survey 2009–10: Final Survey Report.* Denver, CO: Front Range Research Associates, Inc., 2010.

Smith, Phyllis. *Once a Coal Miner: The Story of Colorado's Northern Coal Field.* Boulder, CO: Pruett Publishing Company, 1989.

Stull, James. *Images of America: Erie.* Charleston, SC: Arcadia Publishing Company, 2011.

# INDEX

## S



## U



## V



## W

# ABOUT THE AUTHOR

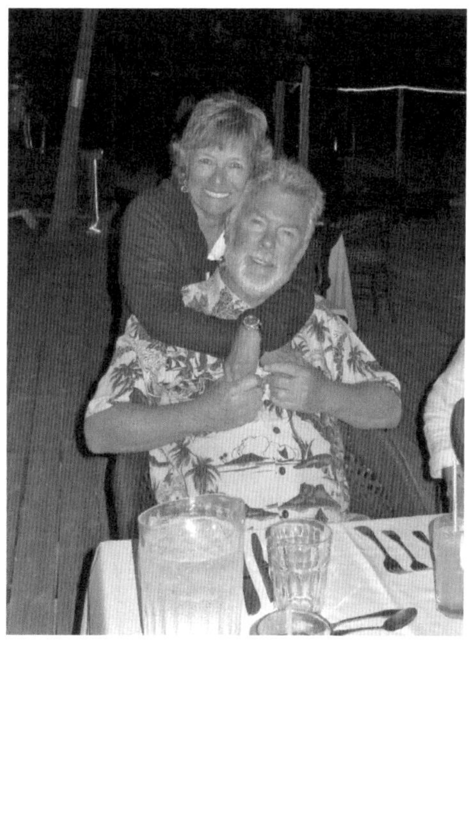

James B. Stull worked in higher education for over forty years and is an emeritus professor at San Jose State University. He earned BA and MA degrees at San Francisco State College and a PhD at Purdue University. He has authored over one hundred publications. He has consulted with more than sixty industrial, government and educational organizations. He is an honorary member of the Erie Historical Society and makes frequent presentations about the history of the town of Erie.